Local Action on Climate Change

There is growing interest in analysing the role and effectiveness of the local scale in responding to the global challenge of climate change. However, while accounts of urban climate change governance are growing, there is now a real need for further conceptual and empirical work to better understand processes of change and uptake across a range of climate change actions.

Local Action on Climate Change examines how local climate change responses are emerging, being operationalised and evaluated within a range of geographical and socio-political contexts across the globe. Focussing on the role and potential of local governments, non-government organisations and community groups in driving transformative change, the authors analyse how local climate change responses have emerged and explore the extent to which they are or have the potential to be innovative or transformative in terms of governance, policy and practice change.

Drawing on a diverse range of case studies, including examples from Vanuatu, Japan, South Africa, Australia, Sweden, the USA and India, this volume will be of great interest to students and scholars of climate change, environmental policy and governance, and sustainability.

Susie Moloney is a senior lecturer in Sustainability and Urban Planning in the School of Global, Urban and Social Studies and Senior Research Fellow in the Centre for Urban Research, RMIT University, Australia.

Hartmut Fünfgeld is an associate professor in Sustainability and Urban Planning in the School of Global, Urban and Social Studies and researcher in the Centre for Urban Research, RMIT University, Australia.

Mikael Granberg is a professor of Political Science, the director of the Centre for Climate and Safety at Karlstad University, Sweden. He is also a research fellow and board member of the Swedish national Centre for Natural Disaster Science, CNDS.

Routledge Advances in Climate Change Research

For a full list of titles in this series, please visit www.routledge.com

Responding to Climate Change in Asian Cities
Governance for a more resilient urban future
Edited by Diane Archer, Sarah Colenbrander and David Dodman

Climate Change Finance and International Law
Alexander Zahar

Urbanization and Climate Co-Benefits
Implementation of win-win interventions in cities
Edited by Christopher Hideo Doll and Jose Puppim de Oliveira

Climate Change and Food Security
Africa and the Caribbean
Edited by Elizabeth Thomas-Hope

Globalizing the Climate
COP21 and the climatisation of global debates
Edited by Stefan C. Aykut, Jean Foyer and Edouard Morena

Adapting Infrastructure to Climate Change
Advancing Decision-Making Under Conditions of Uncertainty
Todd Schenk

Local Action on Climate Change
Opportunities and Constraints
Edited by Susie Moloney, Hartmut Fünfgeld and Mikael Granberg

Local Action on Climate Change

Opportunities and Constraints

Edited by Susie Moloney,
Hartmut Fünfgeld and
Mikael Granberg

Routledge
Taylor & Francis Group
LONDON AND NEW YORK

First published 2018
by Routledge

2 Park Square, Milton Park, Abingdon, Oxfordshire OX14 4RN
52 Vanderbilt Avenue, New York, NY 10017

Routledge is an imprint of the Taylor & Francis Group, an informa business

First issued in paperback 2019

British Library Cataloguing-in-Publication Data
A catalogue record for this book is available from the British Library

Library of Congress Cataloging-in-Publication Data
A catalog record for this book has been requested

ISBN: 978-1-138-68152-1 (hbk)
ISBN: 978-0-367-24883-3 (pbk)

Typeset in Goudy
by Apex CoVantage, LLC

Contents

Tables

Figures

Contributor biographies

Editors

Susie Moloney is Senior Lecturer in Sustainability and Urban Planning and a researcher in the Centre for Urban Research, based in the School of Global, Urban and Social Studies at RMIT University, Melbourne. Her research focuses on land-use planning, urban sustainability and climate change and the implications for policy and governance arrangements particularly at the local and regional scale. She has over ten years' experience working in the public and private sectors on a range of planning and climate change related projects focusing particularly on state and local governments and community organisations.

Dr Hartmut Fünfgeld is an Associate Professor in Sustainability and Urban Planning and a researcher in the Centre for Urban Research, based in the School of Global, Urban and Social Studies at RMIT University, Melbourne. His research focuses on the social and institutional dimensions of local and regional climate change responses, including how to facilitate adaptation to climate change and other socio-ecological trends through strategic planning and participatory processes. Hartmut has over ten years of experience in researching and working on climate change, sustainability and local governance with the public sector and communities in Australia, Europe, Africa and Asia. He holds a PhD in human geography from the University of Heidelberg, Germany.

Dr Mikael Granberg is a Professor of Political Science and the Director of the Centre for Climate & Safety (CCS), Karlstad University, Sweden. He is also research fellow and board member of the national strategic research area, the Centre for Natural Disaster Science (CNDS), Sweden. His main research focus is the handling by society of the challenges from global climate change and if and how institutionalized political practices and norms facilitates or hinders collective action.

Contributors

Kristin Baja is the Climate and Resilience Planner with the Baltimore City, Office of Sustainability. She is responsible for plan development and implementation of the City's resiliency work, which includes climate adaptation, hazard mitigation and climate mitigation efforts. Before joining Baltimore City, Kristin worked for the City of Ann Arbor developing their climate and sustainability framework and plans. Kristin holds a Master of Urban Planning degree and a Master of Science degree from the University of Michigan. In 2016, Kristin was recognized by the White House, during the Obama administration, as a Champion of Change for her work on climate equity and resilience.

Julia Davies is an aspiring young researcher, having recently completed her Masters in Climate Change and Development at the University of Cape Town, in association with the African Climate and Development Initiative. With an academic background in Environmental and Geographical Science, she has a keen interest in the environmental field in general, and in climate change adaptation and mitigation more specifically. Julia has been the lead author in various research projects, including a study related to the development of climate resilient low-income housing and a report to the Water Research Commission, which assesses the applicability of strategic adaptive management in the water services sector.

Dr Christian Dimmer is assistant professor for urban design at the University of Tokyo. As adjunct professor for urban studies he also teaches courses in Sustainable Urbanism, Planning Theory, Theories of Public Space, and Global Urbanism at several leading universities in Tokyo. Christian is partner of the architectural practice Frontoffice Tokyo and co-founder of the civil society organisations Architecture for Humanity Tokyo and Tohoku Planning Forum.

Dr Daniel Kremers is a senior research fellow at the German Institute for Japanese Studies, Tokyo. He received his PhD in Japanese Studies from Martin-Luther-University Halle-Wittenberg in Germany. In his current research he focuses on democratic structures and practices in local communities and their effects on energy policies. Besides that Daniel researches and publishes on Japan's labor migration policies and the global reception of European political philosophy. Together with filmmaker and director Tilman König, Daniel co-produced and co-directed the documentary movie Sour Strawberries – Japan's hidden ‚guest workers'. His recent publications include "Transnational Migrant Advocacy From Japan: Tipping the Scales in the Policy-making Process" in Pacific Affairs, Vol. 87, No. 4 December 2014.

Lars Nyberg is Associate Professor in Risk and Environmental Studies and the research leader of the Centre for Climate & Safety (CCS), Karlstad University, Sweden. His focus is on natural disasters and climate change adaptation with a special interest in how society is impacted and efforts utilised with the aim to reduce risks and mitigate vulnerabilities.

Dr Neha Sami is currently member of faculty at the Indian Institute for Human Settlements in Bangalore, India where she teaches on questions of governance and sustainability as well as anchoring the Research Programme. She holds a Ph.D. in Urban Planning from the University of Michigan, a master's degree in Environmental Management from the Yale School of Forestry and Environmental Studies and a B.A. in Economics from the University of Mumbai. Her dissertation research focused on the political economy of land and governance under conditions of globalization. Prior to beginning graduate school at the University of Michigan, Sami worked with the Boston Redevelopment Authority as an analyst with the Economic Development Division. Sami studies the urban politics of development and governance in post-liberalization India. Her research focuses on the governance arrangements of mega-projects in India. She is also interested in environmental governance questions in Indian cities, particularly around issues of climate change adaptation.

Alexei Trundle is the Future Cities Research Coordinator for the Melbourne Sustainable Society Institute and a doctoral candidate at the Australian-German Climate Energy College, University of Melbourne. He previously worked at the Climate Change Adaptation Programme at RMIT University's Global Cities Research Institute, where he led research projects developing Urban Resilience and Climate Adaptation Plans for Port Vila, Vanuatu and Honiara, Solomon Islands. He is a research associate of the UN Global Compact Cities Programme, and a scientific advisor to UN-Habitat in the Pacific region.

Gina Ziervogel Associate Professor is a geographer by training, with 15 years of experience in the field of adaptation and vulnerability to global environmental change. She is a lecturer in the Department of Environmental and Geographical Science, and a Research Chair at the African Climate and Development Institute at the University of Cape Town. Her research focuses on climate adaptation, transformation and development at both the household and municipal level. She is particularly interested in interdisciplinary and transdisciplinary projects that bring together civil society, government and academics to address these problems collaboratively and creatively. Gina was a lead author on the Intergovernmental Panel on Climate Change (IPCC) Special report on Managing the risk of extreme events and disasters to advance climate change adaptation (SREX) and in 2015 won the South African Young Woman in Science Award.

1 Climate change responses from the global to local scale

An overview

Susie Moloney, Hartmut Fünfgeld and Mikael Granberg

Introduction: scope and aims of the book

If there was any doubt, the Paris Agreement has placed climate change firmly on the agenda of decision-makers and planners at all scales of government (Bulkeley 2015) and also continued to emphasise the need to focus on climate change adaptation. In light of efforts to contain global warming below two degrees by the end of the century, there is a clearer than ever scientific and political imperative to act now in a decisive manner. This entails action on both reducing greenhouse gas emissions as well as planning for the impacts of climate change through adaptation.

The enormity and complexity of the challenge calls for coordinated action, with regard to international cooperation, devising coherent policy frameworks at the national scale and developing local and regional actions and initiatives that involve public and private sectors as well as community groups and households (Harrison and Sundström 2010; Bulkeley and Tuts 2013). At the international level, global institutions such as the United Nations Framework Convention on Climate Change (UNFCCC) provide highly formalised institutional frameworks that help coordinate action across national and otherwise defined administrative boundaries as their core purpose. Under the guidance of the UNFCCC, least-developed nations (LDCs) have developed National Adaptation Programmes of Action (NAPAs) to address their most urgent and immediate adaptation needs, while most governments in the Global North as well as some non-LDCs in the Global South have developed National Adaptation Plans – in many cases these have been cascaded down to the regional level.

Yet to achieve the ambitious goal of retaining global warming to below two degrees, it is increasingly clear that cascading, top-down action from the international to national scale is going to be insufficient. Rather, for climate change action to be catalysed in a way that is needed, transformative change is going to be required, changing aspects of people's everyday lives to changing our governing institutions. Such broad institutional change can only be facilitated with significant degrees of local and regional action. Local governments, community organisations and private companies are already engaging in climate change action, thereby exploring not only new opportunities for innovation (of processes, institutions, products, etc.) and risk mitigation but often, also developing new approaches for collaborating with partners to plan and implement climate change responses. This arena of action provides significant opportunities for catalysing innovative climate change action and achieving more inclusive processes that can lead to effective and widely supported outcomes.

Understanding and situating local climate change action and governance

There is growing interest in better understanding the types of action that are forming in cities and regions around the world to tackle climate change and the value that different forms of local action bring to effectively dealing with the challenges climate change presents. This interest can be traced in the growing body of work that has emerged focusing on the role of local actors in responding to climate change (cf. Connor 2016; Kent 2016; Hoff and Gausset 2016; van Buuren et al. 2016; Salih 2016) and in particular the role of cities, local governments, community organisations and private enterprises involved in 'climate change experiments' (McGuirk et al. 2014; Karvonen et al. 2013; Moloney and Horne 2010; 2015; Bulkeley et al. 2011, 2014; Hoffman 2011; Castan-Broto and Bulkeley 2012). Action, innovation and transformation at the local scale are of interest not only to researchers, but also to activists who are frustrated with the slow progress with multi-lateral treaties, agreements and targets at the international scale, despite recent progress in COP21 in Paris.

While the local level is significant, it is also evident that climate change action requires multi-level governance responses, which coordinate and integrate vertically, across hierarchical levels of governance, as well as horizontally, across different spheres of society (Keskitalo et al. 2016; Moloney and Fünfgeld 2015; Stephenson 2013; Sovacool 2011). Research, policy practice and politics of climate change (climate governance) have for a long time had a strong focus on the global level and the formation of international regimes. Contemporary perspectives, however, have taken greater interest in local and regional dimensions of climate governance and a "more 'polycentric' or multi-level approach [that] regards the plurality of actors and levels and the complexity of their interactions" (Jänicke et al. 2015, p. 3). Increasingly, research investigating the local and regional dimensions of climate governance focuses on aspects such as: examining the actors involved in responding to climate change, on various scales and across scales (Bache et al. 2015; Betsill and Bulkeley 2006); investigating who has formal or informal responsibility and mandates to make decisions and act on climate change (Corfee-Morlot et al. 2011; Craft and Howlett 2013); exploring how climate change 'problems' are framed and 'solutions' identified (Bisaro et al. 2010; Fünfgeld and McEvoy 2011); and uncovering the conditions for enabling and constraining capacities to act in different contexts (Jones et al. 2010; Pahl-Wostl 2009).

Attention in this sphere of analysis also focuses on forms of network governance that includes not only connected government departments and agencies but actors external to administrative bureaucracies such as urban planners, local decision-makers, civil society leaders and private business actors (Luthe et al. 2012; Hanssen et al. 2013; Hajer and Wagenaar 2003). Given the pervasiveness of both cause and effect of anthropogenic climate change, network governance that both complements, and sometimes challenges, traditional forms of government holds potential for developing and implementing innovative and transformative climate change responses. While small, incremental actions can be considered insufficient in the face of systemic changes required, it is in such networked forms of cooperation, collaboration and sometimes conflict where local transformative action is taking place. There is therefore an imperative to better understand the innovative and transformative potential of locally and regionally connected climate change actions, along with the conditions constraining that potential.

When examining any local processes that involves multiple actors and institutions with different capacities and power to make decisions, examining issues of politics and social relations becomes important (Gillard et al. 2015) to better understand the role of particular actors involved in local climate change efforts. Key questions here are: who is included and excluded and why; what are their motives for engagement/dis-engagement; and how do political agendas and goals emerge and how are they enacted through collaborative endeavours? Furthermore, the extent to which climate change is framed as a collective, societal problem is critical for understanding the nature and potential of local responses, including the particular forms of knowledge that are drawn upon and created when devising such responses (Granberg et al. 2016).

In paying attention to issues of politics, social relations, social in-/exclusion and the framing of policy issues and problems, gaining an understanding of power and influence becomes particularly relevant (Beetham 2013). Here, we see power as a wide concept including power of decision and non-decision making (Bachrach and Baratz 1962), power as the manipulation of desires, preferences and worldviews in ways that does not make open coercion necessary (Lukes 1974) and power as a force forming subjects (Foucault 1977; Digeser 1992). This entails that individual (subjects) perceptions of interests, rationality, intentionality, responsibility, the "rules of the game", policy issues and problems, etc., are formed and framed over time in ways that can enable or disable agency. Having access to and exerting power impacts on people's ability to make judgements and choices and on what is perceived as possible (or impossible), desirable (or undesirable) (Bacchi 1999), and – importantly in the context of politicised public discourses on climate change – what "counts as knowledge, what kind of interpretation attains authority as the dominant interpretation" (Flyvbjerg 1998, p. 226). Power, from this perspective, operates in subtle ways impacting the utilisation of knowledge, inclusion and exclusion and societal priorities in general (Bulkeley and Newell 2010, p. 112). Power, therefore, is at the foundation of social practices of politics, planning, religion, etc. (Beetham 2013, p. 3), and of course, this includes social practices relating to climate change responses such as mitigation, adaptation and risk reduction.

With great awareness of how issues of power and conflict manifest themselves at the local scale, we will look at how climate change responses have emerged and are being operationalised and evaluated within a range of geographical and socio-political contexts across the globe. Contributions to this book also illustrate the processes by which dominant logics connected to incumbent policy frames shape or influence the development of local climate change actions and responses. They highlight the relevance of understanding the priorities and politics of the wider local and multi-level policy context, to better understand the conditions for action and how climate change action is formulated, defined and implemented (Granberg et al. 2016). Through these case studies, we examine how climate change is framed and embedded in both discourse and action, and discuss how it can leverage (or constrain) innovations across different policy, social and technical domains. We highlight institutional and political flow-on effects, such as how the broader policy context at the local scale influences the identification, formulation and given priority of climate change action and partially determines the inclusion or exclusion of climate change considerations in policy agendas. Most importantly, the case studies were selected because they lend themselves to a critical examination of the institutional frameworks, actors and capacities that enable and constrain local climate change action.

Accordingly, we envisage this book to feed into the ongoing research and policy dialogue on local climate change responses, through learning from existing practice to advance our theoretical understanding based on lived experience in different empirical contexts. The book's intended target audiences are thus not only researchers and students, but also policy makers and practitioners who work on climate change issues with a keen interest to better understand what drives local climate change action in different parts of the world.

Climate change action context

There is, in principle, scientific consensus that the Earth's climate is changing and that the emergence of global warming over recent centuries is clearly connected to human action and to the development of society (cf IPCC 2014; Oreskes 2004; NAS 2001). The warming of the Earth's atmosphere leads to a diverse range of biophysical impacts entailing rising temperatures, more frequent and more intense downpours, rising sea levels, droughts and in general, an increased frequency of extreme weather events (IPCC 2014). Accordingly, the human challenges arising from climate change develop in a complex interaction between nature, technology and society, and it is clear that a changing climate, among changes to biophysical parameters in the natural environment, also drives demand for changes to social and institutional processes (Glover 2017; IPCC 2014; Granberg and Glover 2014; Bauer et al. 2012). While human and societal action in the face of environmental change is not a new phenomenon per se, contemporary climatic change, through "its speed and scale of impact, combined with the invisibility of causal linkages in everyday life" (Pelling 2011, p. 13) challenges human cognitive abilities to devise solutions in new ways and threatens the survival of whole socio-ecological systems.

Social and institutional change in relation to climate change is likely to require more than just minor adjustments at the policy scale. In a policy sense, climate change responses can roughly be divided into climate change mitigation (reducing greenhouse gas emissions) and adaptation (assessing risks, preparing for and responding to the impacts of climate change) (Keskitalo et al. 2016; Casado-Asensio and Steurer 2014; Granberg and Glover 2014; Granberg and Elander 2007). In one sense, mitigation and adaptation pose vastly different problem structures for policy design and intervention. The effect of reducing greenhouse gas emissions is independent of place – it does not matter where emissions are mitigated as it is the cumulative effect of reducing greenhouse gas emissions into the atmosphere that will slow down climate change. Mitigation therefore requires strong and concerted action at all governance levels and in all geographical settings, but in particular, it calls for coordination of national and international action, e.g., through international agreements and coordinated and cascaded national policy. The United Nations Framework Convention on Climate Change (UNFCCC) process thus far has shown that it is far from straightforward to institutionalise a shared global responsibility, for historical, political and socio-economic reasons. At the local scale, the opportunity for reducing greenhouse gas emissions transcends administrative boundaries, sectors and institutions – from the individual to the multi-national corporations, opportunities for tackling emissions and reducing the carbon footprint can be readily identified using scientific methods. Local action on mitigation therefore can take a myriad of forms, with incentives for action ranging from cost efficiency to improving reputation and managing the risks of future regulation on mitigation.

Adaptation to climate change, on the other hand, is first and foremost a place-based endeavour. In essence, it constitutes a social response to locally and regionally experienced climate change impacts, such as extreme weather events, heatwaves and droughts, changing rainfall patterns and sea level rise. Strong policy efforts at national and international scales are nevertheless required to provide direction and apply principles, such as just and fair adaptation processes and principles for adaptation financing. However, when it comes to planning for future climate change impacts, adaptation action always needs to be locally contextualised and often, customised to local and regional socio-cultural and institutional factors.

However, despite the differences with regard to problem structure, both mitigation and adaptation address a collective action problem without clearly identifiable responsibilities (Keskitalo et al. 2016) even though these problems may be framed as predominantly 'international' (in the case of mitigation) or 'local' (in the case of adaptation). This highlights not only the need for steering by the state but also steering and coordination among broader groups, citizens, businesses and non-governmental organisations. Responsibilities for climate change action are unclear and thus need to be negotiated between citizens, private businesses and government entities. Both mitigation and adaptation involve considerable uncertainties around what measures and investments are needed today to avoid damage tomorrow and who ought to take responsibility for those actions.

When considering the dichotomy of climate change mitigation and adaptation, it is important to acknowledge that the latter is closely linked in policy and practice as well as conceptual terms to disaster risk reduction (DRR). DRR is primarily about preventative or remedial action stemming from concrete catastrophic events even though the reduction aspect opens up for more long-term perspectives (Birkmann 2013; Birkmann and von Teichman 2010). DRR policy interventions are frequently framed using the concept of resilience. Resilience has many possible interpretations but most commonly refers to a system's ability "to resist, absorb, accommodate to and recover from the effects of a hazard in a timely and efficient manner, including through the preservation and restoration of its essential basic structures and functions" (UNISDR 2009, p. 24) when exposed to hazards. This conservative notion of resilience being mainly about 'bouncing back' and maintaining a status quo despite disruption has been very pervasive in the DRR literature and policy, though gradually the transformative potentials of resilience are highlighted in policy interventions labelled as "build back better" (UNISDR 2015) or "bounce forward" (The Kresge Foundation 2015). This draws attention to broader social and institutional changes and the need to address social equity and economic justice.

To help frame the scope and content of this book, it is useful to outline the range of actions and measures most commonly associated with mitigation and adaptation and relate them to the scales and actors typically involved. The discussion thus far and the overview of typical actions contained in Table 1.1 both frame and illustrate the breadth of climate change responses at the local scale and the range of actors involved.

Interpreting local climate change action: conditions enabling and constraining innovation and transformation

Local governments and other actors operating at the local scale increasingly invest effort and resources in devising localised climate change responses that fit with the broader climate governance and political context (Bulkeley and Castán Broto 2013, Rosenzweig et al. 2010, 2011). A critical question that this book seeks to address, following Pelling (2011), is: what can we learn from local initiatives on mitigation and adaptation that provide opportunities not only for incremental action but also for transformative change? We take the (arguably normative) view that transformation will not only be desirable but necessary for societies to deal with the challenges posed by climate change and keeping their ability to thrive into the future. We therefore consider it a core task of social science research to identify and critically examine opportunities for transformative action at the local scale and thus contribute to enabling more individuals and organisations to change the way they operate and as an aggregate outcome of such individual changes, how our societies function.

Analysing local climate change responses, however, is not a straightforward task. Only some local actions have universal relevance, and the majority of mitigation and, even more so, adaptation measures are necessarily highly customised to local context (Bulkeley and Castán Broto 2013). Context specificity, including

Table 1.1 Multi-level climate change actions and measures

Actions and measures	*Range of scale(s) of application (largest to smallest)*	*Main actors involved*	*Example*
Mitigation actions (to reduce GHG emissions)			
• Carbon tax and Emissions Trading Schemes	Supranational – regional	Supranational institutions, Governments	European Union Emissions Trading System (EU ETS)
• Reducing/removing fossil fuel subsidies	National – regional	Governments	Changes in the tax system, 'green' tax reform
• Carbon sequestration and carbon storage measures	Regional – local	Governments, private sector, land owners	Carbon Farming Initiative (CFI), Australia
• Incentivising and investing in renewable energy	Supranational – local	Supranational institutions, governments, private sector, civil society	Renewable Energy Law (EEG), Germany
• Energy efficiency measures	National – local	Governments, peak bodies, professional associations	Building and appliance design standards, insulation
• Changes to urban form and density	Regional – local	Planning authorities, governments	Land use zoning to reduce low density car based development, building design, greening cities
• Increasing sustainable transportation and mobility	National – local	Planning and Transport authorities, governments, civil society groups, households	Investment in public transport, urban design to encourage walking and cycling
• Reducing energy consumption through changing everyday practices	Regional – local	Governments, civil society groups, households	National or local government engagement campaigns Changes in social practices implicated in energy and water consumption

(Continued)

Table 1.1 (Continued)

Actions and measures	*Range of scale(s) of application (largest to smallest)*	*Main actors involved*	*Example*
Adaptation actions (reducing vulnerabilities, risks, building adaptive capacities)			
• Hard infrastructure measures to adapt to climate change impacts	Regional – local	Governments, planning authorities, engineering and construction sectors	Building sea walls and dikes
• Land use based adaptation	Regional – local	Governments, planning authorities	Restricting particular forms of land use, such as industrial use, due to climate change risks (e.g., inundation)
• Conducting risk and vulnerability assessments	Regional – local	Governments, private businesses	Mapping of risks Developing local and regional risk scenarios
• Building adaptive capacity	Regional – local	Governments, private businesses, civil society groups	Building social capital and institutional adaptive capacity through training, informal learning and exchange
• Increasing food security and access to food	Regional – local	Civil society groups, governments	Supporting local agricultural production or urban farming
• Assessing and managing health vulnerabilities	Regional – local	Civil society groups, governments	Managing heat stress amongst elderly people

geographical, social, economic and political conditions and contingencies, is one of the critical success factors for climate change adaptation at the local scale (Moser et al. 2014), making it difficult to single out and generalise activities and initiatives that promise to be equally successful (i.e., of equal likelihood to result in transformative outcomes) when transferred to other locations.

This highlights epistemological and methodological difficulties in creating a broader understanding of what constitutes transformative change at the local scale. Central to this challenge is how to assess and compare local scale transformative action, as transformation in one context does not necessarily lead to transformation, or would not even be considered transformative, in another context (Pelling 2011). On what basis then can we assess the potential or likelihood of transformative change in different realms and contexts? The importance of context specificities can also be identified in the multitude of different approaches to analysing climate change actions. These approaches draw on a wide range of disciplines and literature from research strands that include climate change governance, socio-ecological systems, socio-technical systems, urban climate governance, progressive place governance, social innovation, theories of practice, institutional adaptive capacity building and more.

Accordingly, transformational systemic change may be difficult to discern, but empirically identifying specific emergent processes and types of transformative change at various scales and in different contexts is possible. Examining local climate change action thus provides us with an opportunity to identify and understand *transformative processes* – rather than identifying generic methods leading to specified results. Studying local climate change action affords the opportunity to identify possible points of intervention in – albeit amorphous – planning and decision-making processes and highlights how initiatives of various kinds are mobilising, or have the potential to mobilise, transformational change across a range of social, infrastructural and networked and collaborative local and regional governance systems and contexts. Table 1.2 provides an overview of conditions that can enable or constrain innovative and transformative climate change action at the local scale. These conditions are evident to varying degrees in the case studies presented in this book, including typical local actions, examples of local collaboration and potentials for transformation. While the conditions are not the only success criteria for engendering innovative and transformative climate change action, they provide a conceptual framework for reflecting on and interpreting the case studies that follow this chapter.

Overview of chapters

Here we provide a short introduction to each case study chapter guided by three questions: *What* is the chapter about, *why* is the chapter interesting and *what* lessons can we learn from the chapter? In the concluding chapter, we will reflect on the conditions enabling and constraining innovative or transformative action in each context offering some directions for future research, policy and action.

Table 1.2 Conditions enabling and constraining local innovative and transformative action on climate change mitigation and adaptation

Condition	*Institutional and organisational dimensions in which condition applies*	*Enabling condition*	*Constraining condition*
Clarity of roles and responsibilities	Multi-level governance of climate change Intra-organisational division of roles and responsibilities	Efficient and effective planning and decision-making across multiple scales and institutional levels	Confusion about roles and responsibilities leading to stalling and inaction Conflicting actions increasing emissions and maladaptation
Political commitment and high-level support	Democratic decision-making processes Organisational buy-in	Political commitment demonstrates importance and enables buy in from across the organisation/ constituency	Lack of political commitment leads to neglect or active discouragement of climate change action
Organisational and institutional co-ordination and integration	Horizontal and vertical collaboration across institutions and departments	Well co-ordinated climate change action leads to mainstreaming into all organisational processes	Poor coordination leads to sporadic and fragmented action
Problem/policy framing and formulation	Multi-level/actor participatory approach Value based framing and prioritisation in the formulation of issues on the policy agenda	Climate change is framed as an all-encompassing problem that requires coordinated and sustained action	Climate change action remains a niche task of environment departments
Local context and specific circumstances	Translation of climate challenges/ problems in relation to local contextual conditions Inclusion of multiple local scale actors and forms of knowledge in decision making	Climate change action is tailored to and grounded in locally specific circumstances, drawing on local knowledge	Climate change action ignores local knowledge and context and is devised in a top-down fashion
Social learning and Experimenting	Developing capacity through breaking path dependency Commitment to innovation, reflexive learning and developing new pathways of action	Experimenting and social learning is encouraged, leading to innovation and transformative action	Experimentation is discouraged or not resourced and climate change action is required to fit into 'business as usual'

In Chapter 2, *Collaborative governance: a means for integrating climate change action in Australian local government?*, Hartmut Fünfgeld and Susie Moloney look at capacity building and innovative organisational processes focusing on risk management and climate change adaptation in the Australian cities of Geelong and Melbourne. In the former, the city of Geelong demonstrates a proactive and progressive approach to internally coordinating a response to climate change through the development of innovative decision-making tools and processes. In the latter, the work of local government climate change alliances in Melbourne demonstrates the potential for regional scale planning and collaboration to increase the capacity of local actors and communities to respond to climate change. This is an interesting case as the weak local government context in Australia creates conditions where local governments are compelled to address climate change challenges through the formation of different types of partnerships trialling innovative practices and attempting to push the frontiers of the whole chain of climate change adaptation practices (process, organisation and implementation beyond risk assessment). The Geelong local government aimed at innovating decision-making processes but struggled with handling complexity. The local government climate change alliances seek to shape and manage a range of practices and navigate the complexities of doing so across jurisdictional, organisational and institutional boundaries. The chapter highlights the opportunities and challenges of network governance and capacity building by deepening our understanding of the challenges involved in 'doing' collaboration through informal/formal governance processes focusing on climate change planning and mainstreaming into local government practices.

In chapter 3, *Governance and agency beyond boundaries: climate resilience in Port Vila's peri-urban settlements*, Alexie Trundle studies urbanisation processes and climate resilience in the peri-urban areas of Port Vila, Vanuatu. He does this through an historical approach using a political economy lens observing urban development processes in the intersection of urbanisation and climate change impacts and risks. This is a vital case study highlighting the constraints to effective urban governance in general, and with regard to climate change adaptation in particular. It looks at efforts to include and empower indigenous communities and also on conditions for social learning and inclusion of experiential (non-expert) knowledge in local climate change action. The chapter emphasises the challenges/risks in the exclusion of informal spaces and settlers from institutional/formal urban processes and focuses on the need for an in-depth understanding of a particular location's historical and cultural origins and of the local non-institutional arrangements that maintain or sustain them. In line with this, the chapter also shows, through its long-term horizon, that arbitrary urban boundaries can limit effective city development when handling cross-boundary issues such as climate change adaptation, climate resilience and connected vulnerabilities.

In chapter 4, *'Learning by doing': lessons from the co-production of three South African municipal climate change adaptation plans*, Julia Davies and Gina Ziervogel look at the production processes of climate change adaptation plans in three local settings in the Western Cape Province of South Africa. This is an important case study of climate change adaptation as a relative novelty in a context that lacks readily transferrable frameworks to guide adaptation planning. Accordingly, this is a study of local governments that have largely adopted a 'learning-by-doing' approach. From the three case studies in this chapter, it is clear that there is no 'one-size-fits-all' approach to local climate change adaptation. It also shows that political will and strong support from senior public officials for the adaptation agenda paired with a collaborative and integrated approach is key for enabling more effective policy processes. It is evident that more immediate social and economic issues pose a challenge for prioritising and mainstreaming climate change adaptation and that the positioning of climate change as an issue within organisational arrangements influences the capacity for transformation.

In chapter 5, *Facilitators and barriers: regional diversity among local climate protection projects and regional revitalisation efforts in rural Japan*, Christian Dimmer and Daniel Kremers focus on two major socio-economic challenges – rapid demographic ageing and a high level of energy dependency and insecurity. They look at how these challenges are linked to the prosperity of local communities in two cases of local climate action in rural Japan. Studies of climate change action in peripheral settings are less common, and the connection between rapid demographic ageing and climate change action has had limited analysis. This chapter presents some of the complexities of intersecting policy problems in shaping local climate change action in Japan and is valuable in contributing new perspectives and new knowledge to studies of climate change action. From the two case studies discussed in this chapter, we learn that ambitious objectives formulated by higher-tier government can run into structural problems, hindering implementation at the local level when municipalities lack vision, knowledge or capacities to implement national objectives. In order to accelerate local climate action, there is a need to frame the issue of climate change in a broader perspective emphasising the co-benefits of local action by focusing on the potential to improve broader (economic, social, etc.) local conditions in ways that also increase the adaptive capacity and resilience of communities.

In chapter 6, *Multi-level climate change planning: scale, capacity, and the authority for local action*, Neha Sami examines local climate change action in two Indian cities, Bangalore and Chennai, focusing explicitly on urban settlements and their ability to respond to climate-related challenges. The focus in these case studies on urban settlements in the Indian context is of particular interest considering the scale and speed of India's urban transition with more than half the country defined as urban by 2050. India's urban populations are increasingly vulnerable to a range of environmental and socio-economic risks that, potentially, will be exacerbated by climate change. A key activity identified by the Indian cases is that of urban planning as a vehicle for climate change action and mitigating impacts. Both examples illustrate that coordination across different agencies and scales are important but challenging in the Indian context. Responsibility for planning and governance functions are typically distributed across a range of city, parastatal and state-level agencies, and if the power over and control of urban planning is not clearly defined, it may lack in operational capacity, reducing both frequency and efficiency in local climate action.

In chapter 7, *Climate change adaptation, city competitiveness and urban planning in the city of Karlstad, Sweden*, Mikael Granberg and Lars Nyberg study how municipal climate change actions are mediated through dominant policy priorities shaping city development goals and processes through a case study of the city of Karlstad, Sweden. Waterfront redevelopment, as an integral component in efforts to make cities more attractive and competitive, is common in many cities around the world. Studying the connection between this trend, climate risks and climate change adaptation is therefore important for research on local climate change action. One conclusion from this case study is that there is a growing awareness of the double challenge for planning and city redevelopment as the city develops waterfront housing. The studied city is active and innovative in implementing policies and utilising different actors to handle climate-related risks. However, the focus on competitiveness is creating a frame within which formulation, prioritisation and climate actions are defined. Accordingly, even in a context like Sweden with its profile as a leader in climate change action, the dominance of a growth imperative continues to outweigh the potential impacts of climate change risks.

In chapter 8, *Climate Resilience: Transformative Efforts in Baltimore City*, Kristin Baja and Mikael Granberg look at local government climate action in terms of efforts to address equity issues through proactive planning when working on risk reduction and preparedness in the city of Baltimore in the US. Baltimore's urban center is an important asset that also faces immense social and economic challenges. The city's economic growth objectives are also centered on waterfront development and redevelopment, including industrial, commercial, recreational and residential development. The city of Baltimore has a long history of deliberate segregation which underpin many of the economic and social challenges the city faces today. Accordingly, the city's combination of shocks and stresses cuts across social, economic and environmental factors. Studying local climate action in a setting confronted with significant social tensions is valuable, and accordingly, this case study makes an important contribution to our knowledge around creating climate resilience in a complex social setting. Striving for climate resilience in a segregated city with social and economic tensions necessitates a focus on equity and consideration of those areas and communities most vulnerable. Equity goals need to be integrated into planning policies and processes and not treated as separate to other development goals. This case study highlights how socio-economic diversity, historic redlining (racial segregation), lack of resources and accessibility are all implicated in creating a climate-resilient city. The Baltimore case also highlights the committed, proactive and innovative work of local government in the face of profound challenges involved in addressing social, economic and environmental impacts in a context where communities of color and low-income residents are disproportionately affected by climate change.

In the concluding chapter, we draw out key lessons from each chapter and identify thematic similarities and differences across the cases. In doing so, we reflect on the conditions enabling and constraining local climate change actions identified in Table 1.2 and highlight where opportunities for transformative action are emerging or have the potential to emerge. While these potential openings or opportunities are identified, we recognise that the transformative potential of local climate change action is highly contingent and susceptible to wider changes at regional, national and global scales. Despite the ongoing dynamic and uncertain context within which local actors come together to respond to climate change, we can discern some useful lessons from the case studies presented in this volume situated in different contexts and continents across the world. At the very least, these cases provide fascinating stories of different local imperatives and experiences in responding to climate change, while at the same time highlighting the shared challenges and opportunities for innovation and transformation emerging through local climate change action. We finish with some conclusions around emerging trends, gaps and needs for future research to better enable local transformative change.

References

Bacchi, C.L. (1999). *Women Policy and Politics: The Construction of Policy Problems*. London: Sage.

Bache, I., Bartle, I., Flinders, M., and Marsden, G. (2015). Blame games and climate change: Accountability, multi-level governance and carbon management. *The British Journal of Politics and International Relations*, 17(1), 64–88.

Bachrach, P., and Baratz, M.S. (1962). The two faces of power. *American Political Science Review*, 56, 947–952.

Bauer, A., Feichtinger, J., and Steurer, R. (2012). The governance of climate change adaptation in 10 OECD countries: Challenges and approaches. *Journal of Environmental Policy & Planning*, 14(3), 297–304.

Beetham, D. (2013). *The Legitimation of Power*. 2nd ed. Basingstoke: Palgrave Macmillan.

Betsill, M.M., and Bulkeley, H. (2006). Cities and the multilevel governance of global climate change. *Global Governance*, 12(2), 141–159.

Birkmann, J. (2013). *Measuring Vulnerability to Natural Hazards – Towards Disaster Resilient Societies*. 2nd ed. New York: United Nations University Press.

Birkmann, J., and Teichman, K. (2010), Integrating disaster risk reduction and climate change adaptation: Key challenges – scales, knowledge, and norms. *Sustainability Science*, 5(2), 171–184.

Bulkeley, H. (2015). Can cities realise their climate potential? Reflections on COP21 Paris and beyond. *Local Environment*, 20(11), 1405–1409.

Bulkeley, H., and Betsill, M.M. (2003). *Cities and Climate Change: Urban Sustainability and Global Environmental Governance*. New York: Routledge.

Bulkeley, H., and Castán-Broto, V. (2013). Government by experiment? Global cities and the governing of climate change. *Transactions of the Institute of British Geographers*, 38(3), 361–375.

Bulkeley, H., Castán Broto, V., Hodson, M., and Marvin, S. (2011). *Cities and Low Carbon Transitions*. New York: Routledge.

Bulkeley, H., Castán Broto, V., and Maassen, A. (2014). Low-carbon transitions and the reconfiguration of urban infrastructure. *Urban Studies*, 51, 1471–1486.

Bulkeley, H., and Newell, P. (2010). *Governing Climate Change*. London: Routledge.

Bulkeley, H., and Tuts, R. (2013). Understanding urban vulnerability, adaptation and resilience in the context of climate change. *Local Environment*, 18(6), 646–662.

Casado-Asensio, J., and Steurer, R. (2014). Integrated strategies on sustainable development, climate change mitigation and adaptation in Western Europe: Communication rather than coordination. *Journal of Public Policy*, 34, 437–473.

Castán-Broto, V., and Bulkeley, H. (2012). A survey of urban climate change experiments in 100 cities. *Global Environmental Change*, 23(1), 92–102.

Connor, L.H. (2016). *Climate Change and the Anthropos: Planet, People and Places*. London: Routledge.

Digeser, P. (1992). The fourth face of power. *The Journal of Politics*, 54(4), 977–1007.

Flyvbjerg, B. (1998). *Rationality and Power. Democracy in Practice*. Chicago: University of Chicago Press.

Foucault, M. (1977). *Discipline and Punish: The Birth of the Prison*. New York: Random House.

Gillard, R., Gouldson, A., Paavola, J., and Van Alstine, J. (2015). Transformational responses to climate change: Beyond a systems perspective of social change in mitigation and adaptation. *WIREs Climate Change Review*, 7, 251–265, doi:10.1002/wcc.384

Glover, L. (2017). *Community-Owned Transport*. London and New York: Routledge.

Granberg, M., and Elander, I. (2007). Local governance and climate change: Reflections on the Swedish experience. *Local Environment*, 12(5), 537–548.

Granberg, M., and Glover, L. (2014). Adaptation and maladaptation in Australian national climate change policy. *Journal of Environmental Policy and Planning*, 16(2), 147–159.

Granberg, M., Nyberg, L., and Modh, L-E. (2016). Understanding the local policy context of risk management: Competitiveness and adaptation to climate risks in the city of Karlstad. *Sweden. Risk Management*, 18(1), 26–46.

Hanssen, G.S., Mydske, P.K., and Dahle, E. (2013). Multi-level coordination of climate change adaptation: by national hierarchical steering or by regional network governance? Local Environment, 18(8), 869–887. Retrieved May 31, 2014, from www.tandfonline.com/doi/abs/10.1080/.U4-uCPQW2So

Harrison, K., and Sundstrom, L.M. (2010). Introduction: Global commons, domestic decision. In: Harrison, K. and Sundstrom, L. (eds.) *Global Commons, Domestic Decision: The Comparative Politics of Climate Change*. Cambridge: MIT Press, 1–23.

Hoff, J., and Gausset, Q. (eds.) (2016). *Community Governance and Citizen-Driven Initiatives in Climate Change Mitigation*. London: Routledge.

Hoffman, M.J. (2011). *Climate Governance at the Crossroads: Experimenting with a Global Response*. New York: Oxford University Press.

IPCC (2014). *Climate Change 2014: Impacts, Adaptation, and Vulnerability*. Cambridge: Cambridge University Press.

Karvonen, A., Evans, J., and Van Heur, B. (2013). The politics of urban experiment's. In: Hodson, M. and Marvin, S. (eds.) *After Sustainable Cities?* London: Routledge, 104–115.

Kent, J. (2016). *Community Action and Climate Change*. London: Routledge.

Keskitalo, E., Carina, H., Juhola, S., Baron, N., Fyhn, H., and Klein, J. (2016). Implementing local climate change adaptation and mitigation actions: The role of various policy instruments in a multi-level governance context. *Climate*, 4(1), 7; 1–11, doi:10.3390/cli4010007.

The Kresge Foundation (2015). *Bounce Forward: Urban Resilience in the Era of Climate Change*. Troy: The Kresge Foundation.

Lukes, S. (1974). *Power: A Radical View*. Houndmills: Palgrave Macmillan.

Lukes, S. (2005). *Power: A Radical View*. 2nd ed. Houndmills: Palgrave Macmillan.

Lundqvist, L.J., and Biel, A. (2007). From Kyoto to the town hall: Transforming national strategies into local and individual action. In: Lundqvist, L.J. and Biel, A. (eds.) *From Kyoto to the Town Hall: Making International and National Climate Policy Work at the Local Level*. London: Earthscan, 12–20.

McGuirk, P.M., Robyn, D., and Bulkeley, H. (2014). Repositioning urban governments? Energy efficiency and Australia's changing climate and energy governance regimes. *Urban Studies*, 51, 2717–2734.

Moloney, S., and Fünfgeld, H. (2015). Emergent processes of adaptive capacity building: Local government climate change alliances and networks in Melbourne. *Urban Climate*, 14, 30–40.

Moloney, S., and Horne, R. (2015). Low carbon urban transitioning: From local experimentation to urban transformation? *Sustainability* (Switzerland), 7, 2437–2453.

Moloney, S., Horne, R.E., and Fien, J. (2010). Transitioning to low carbon communities – From behaviour change to systemic change: Lessons from Australia. *Energy Policy*, December, 38(12), 7614–7623.

NAS (National Academy of Sciences) (2001). *Climate Change Science: An Analysis of Some Key Questions*. Washington, DC: National Academy Press.

Oliver-Smith, A., Alcántara-Ayala, I., Burton, I., and Lavell, A. (2016). *Forensic Investigations of Disasters (FORIN): A Conceptual Framework and Guide to Research*. IRDR FORIN Publication No.2. Beijing: IRDR.

Oreskes, N. (2004). The scientific consensus on climate change. *Science*, 306(5702), 1686.

Pelling, M. (2011). *Adaptation to Climate Change: From Resilience to Transformation*. London: Routledge.

Rosenzweig, C., William, S., Stephen, A.H., and Shagun, M. (2011). *Climate Change and Cities: First Assessment Report of the Urban Climate Change Research Network*. Cambridge: Cambridge University Press.

Rosenzweig, C., William, S., Stephen, A.S., and Mehrotra, S. (2010). Cities lead the way in climate-change action. *Nature*, 467(7318), 909–911.

Salih, M.A. (ed.) (2016). *Local Climate Change and Society*. London: Routledge.

Sovacool, B.K. (2011). An international comparison of four polycentric approaches to climate and energy governance. *Energy Policy*, 39(6), 3832–3844.

Stephenson, P. (2013). Twenty years of multi-level governance: "Where does it come from? What is it? Where is it going?. *Journal of European Public Policy*, 20, 817–837.

Storbjörk, S. (2007). Governing climate adaptation in the local arena: Challenges of risk management and planning in Sweden. *Local Environment*, 12(5), 457–469.

Storbjörk, S. (2010). 'It takes more to get a ship to change course': Barriers for organizational learning and local climate adaptation in Sweden. *Journal of Environmental Policy & Planning*, 12(3), 235–254.

UNISDR (2009). *UNISDR Terminology on Disaster Risk Reduction*. Geneva: United Nations.

UNISDR (2015). *Sendai Framework for Disaster Risk Reduction 2015–2013*. Geneva: United Nations.

van den Berg, M., and Coenen, F. (2012). Integrating climate change adaptation into Dutch local policies and the role of contextual factors. *Local Environment*, 17(4), 441–460.

van Buuren, A., Eshuis, J., and van Vliet, M. (eds.) (2016). *Action Research for Climate Change Adaptation: Developing and Applying Knowledge for Governance*. London: Routledge.

2 Using collaborative governing to catalyse climate change action in Australian municipalities

Hartmut Fünfgeld and Susie Moloney

Introduction

Complex policy challenges, such as responding to climate change at the local scale, are difficult to address comprehensively and efficiently only with established planning and decision-making models of "plan-do-check-act" (O'Brien et al. 2013). The multilayered complexity of climate change and the institutional challenges involved in responding to its impacts are a poor fit for using straightforward, top-down planning and decision-making models. Climate change and other global megatrends, such as environmental degradation, urbanisation and the rise of the middle class and related consumption patterns, call for integrated policy responses that cut across existing administrative siloes and hierarchies of decision-making. In response to inherently social challenges, such policies need to involve actors outside of government, who work together with government to identify and frame problems and devise effective and politically and socially acceptable solutions.

The necessity for changes to existing policy-making paradigms is reflected in an increasing awareness of networked and collaborative forms of governance, where state authority, while important, is just one of many actors involved in the problem framing and decision-making process at the science-policy interface. With regard to climate change, collaborative and networked forms of governance have been heralded as institutional fixes that have the potential to bring to the surface innovative policy responses to climate change that are fit for purpose. At the same time, such connected forms of governance are also a way of challenging the status quo and provide an opportunity for fundamentally transforming institutional structures and processes to better enable them to cope with the complex challenges of climate change (Kinnear et al. 2013; Gash 2016).

In this chapter, we critically examine the role of different forms of collaboration in local planning and decision-making for local climate change mitigation and adaptation. We look into networked forms of governance in real-world contexts by focussing on intra-organisational and inter-municipal collaboration as two specific situations where networked modes of governing can be observed. Both case studies are located in the state of Victoria, Australia and cover two scales of climate change adaptation planning and decision-making. The first focuses on the adaptation planning work of the city of Greater Geelong, a large regional city authority located a short distance from Melbourne. The second case examines the adaptation work of a 'climate change alliance' – a formalised collaborative of a group of local governments encompassing Melbourne's western metropolitan region (including the city of Greater Geelong). Drawing on the findings of recent and ongoing empirical research in both contexts, we discuss the relevance and accuracy of conceptual models for collaborative forms of local climate change governance from a practical perspective. In critically analysing existing notions of the role and potential of collaboration, we also seek to understand the facets of networked governance most pertinent to climate change action at the local scale, by highlighting institutional and political success factors and constraints for such governance arrangements.

Promises of collaborative governance for local-scale climate change action

Collaborative forms of governance promise to provide more inclusive and more creative avenues for tackling complex societal challenges, such as climate change. By relying on polycentric, non-hierarchical institutional arrangements, collaborative forms of governance are heralded as better suited to policy and decision-making in a hyper-connected world with high degrees of complexity (Newman et al. 2004). However, collaborative forms of governance add institutional complexity that can create additional challenges to the effectiveness and, in particular, the efficiency of planning and decision-making processes. To better understand what collaborative governance approaches can offer for facilitating local climate change action, it is thus useful to characterise these approaches and consider how they can enhance existing climate change planning and responses.

In colloquial use, the term 'governance' is often referred to as an alternative to hierarchical notions of 'government'. Governance implies a degree of collaboration among different actors – typically between governmental and non-governmental actors, seemingly making 'collaborative governance' as a further conceptual refinement obsolete. Most scholars, however, apply the term collaborative governance in more refined ways (Gash 2016).

At times, the term collaborative governance is defined to describe a range of decision-making activities where policy (or rules, regulations and practices) is produced through interaction and communication among partners representing a diversity of interests (Conley and Moote 2003). Some definitions focus on the fact that state and non-state stakeholders engage in consensus-orientated deliberative processes (Johnston et al. 2011), while others emphasise more specific procedural aspects of policy-making processes as unique to collaborative governance, such as a focus on 'interest aggregation', i.e., combining perspectives, resources and skills of a group of people or organisations (Lasker and Weiss 2003). Others yet again highlight that collaborative governance is uniquely focused on resolving shared policy problems and dilemmas (Gerlak and Heikkila 2006).

More specific definitions narrow in on the role of public organisations that involve themselves in a set of activities to co-produce shared goals and strategies (e.g., Davies and White 2012; Koontz and Thomas 2006). Here, the emphasis is on different forms of collaboration between public organisations, which includes joint goal-setting, decision-making and implementation. This notion highlights that collaborative governance is not only about collaboration in a process but essentially, about sharing positive power to act on collective problems among partners (Taylor and de Loë 2012).

Involvement in collaborative forms of governance is voluntary. Because participating partners need to be able to see the benefit of the collective endeavour to their own organisation or to themselves, any collaborative action needs to be based on principles of procedural fairness that involves transparent decision-making processes, which in turn legitimise the collaborative endeavour. Fairness and transparency are considered the 'bread and butter of collaborative governance' (Gash 2016, p. 457). To safeguard transparency and legitimacy, collaborative governance requires strong yet consultative leadership – so-called 'facilitative leaders' who have an ability and desire to carefully manage the relationships within the collaborative (Bidwell and Ryan 2006).

Collaborative forms of governance can be identified at multiple jurisdictional and administrative levels, from the local to the international, and across public and private sectors involved in addressing complex policy problems. Frequently, collaborative governance approaches operate side by side, or are overlaid with, more traditional forms of governance led by government, with individual organisations operating in their hierarchically organised 'home' structures while at the same time engaging in collaborative forms of governance, e.g., as part of a specific project or initiative (Newman et al. 2004).

Collaborative governance comes with significant challenges that can make it difficult to implement effectively in practice. Partners in collective governance regimes need to not only have shared goals but also develop mutual trust and good co-working practices, which act as the glue for collaboration and are important ingredients for generating and maintaining legitimacy. This is critical given the voluntary nature of the endeavour, where members of the collaborative can quit if unsatisfied with deliberative processes or if collective goals no longer align with their own. As collaborative governance arrangements typically emerge in response to complex policy problems, this also means collaborative effort needs to navigate institutional and political complexity in order to be successful. In addition, collaborative governance endeavours come with inherently tentative administrative systems and processes, which still need to mature in the process of progressing the collaborative effort. These can place further obstacles in the way of successful collaboration (Gash 2016). However, little empirical evidence exists with regard to how these obstacles play out in real-world contexts and how they are overcome.

In the context of collaborative local climate change action, collaborative governance is not only a widely proposed approach for facilitating practical collaboration; it also provides a useful analytical lens for examining collaboration within and across organsiations. Local and regional collaboration on climate change mitigation and adaptation typically features many characteristics of collaborative governance. The most productive collaboratives on climate change action involve actors from governmental and civil society sectors, who come to the process with a diverse set of interests, ranging from altruistic goals, such as leaving behind a better place for future generations, to externally imposed aims (e.g., compliance with carbon emissions or risk-management regulations), to corporate social responsibility aims that are embedded in ulterior goals of profit maximisation (Linnenluecke and Griffiths 2013). Local climate change collaboratives, therefore, can provide interesting insights into processes of attempted societal convergence towards a shared goal (e.g., reducing greenhouse gas emissions, or reducing the risk arising from climate change impacts), which – in times of divisive climate change identity politics – makes them important objects of scientific social research.

Different modes of governing at local and regional scales

Climate change mitigation and adaptation have entered the strategic plans of municipalities and the agendas of civil society organisations in myriads of ways – from in-house environmental sustainability strategies to sophisticated, software-based solutions for identifying and monitoring climate change impacts and risks on operational systems. Classifying and interpreting 'how local climate change action is governed' is thus far from clear-cut, even when focussing solely on the comparatively simple hierarchical administrative structures found in the local government sector.

Here, Bulkeley and Kern's (Bulkeley and Kern 2006) discussion on *multiple modes of governing*, based on a comparative analysis of climate change policy in Germany and the UK, provides a useful point of departure. Deliberately distancing themselves from a binary view of *government versus governance*, Bulkeley and Kern identified four specific forms of governing that are prevalent at the municipal scale (ibid.: 2242):

1 *Self-governing:* the capacity of local government to govern its own activities, which relies on processes of organisational management. Examples include energy efficiency schemes within buildings and purchasing green energy.
2 *Governing by provision*: the shaping of practice through the delivery of particular forms of service and resource, accomplished through practical, material and infrastructural means. Examples are the creation of public transport service providers or energy service providers owned or co-owned by municipalities.
3 *Governing by authority:* the use of traditional forms of authority, such as regulation and direction which persist despite reforms. This takes place through the use of sanction. Examples are strategic planning to enhance energy conservation and pedestrianisation.
4 *Governing through enabling:* the role of local government in facilitating, co-ordinating and encouraging action through partnership and private and voluntary sector agencies, and to various forms of community engagement (through persuasion, argument and incentives). Examples include education campaigns on transport alternatives or providing grants for energy efficiency measures.

In their analysis of climate action in Germany and the UK, the two authors found that the majority of measures undertaken in relation to four principal spheres of climate change action (i.e., energy, transport, urban planning, and waste management) were predominately in the energy sector and concentrated in the self-governing mode (ibid.). They assessed modes of governing across each sphere of action and found that a significant proportion of local climate protection policies involved *governing by enabling*, which includes promotion activities, partnership development and providing financial incentives or subsidies to encourage action by other actors. This mode of governing also included shaping policy goals and the delivery of infrastructures and services in partnership with other actors. In short, their review of both Germany and the UK revealed that most municipal responses focused around self-governing and enabling modes of governing.

Rather than focusing on the need to improve the evidence base for action and the regulatory capacity of municipalities (two of the frequently referred to barriers that seem to hold back climate change action, cf. Bulkeley and Kern 2006; Allman et al. 2004; Demerit and Langdon 2004), shifting emphasis from provisioning and authority as modes of governing to self-governing and enabling modes can provide an effective means for demonstrating action in the short run – and in the absence of strong regulatory policy frameworks at higher levels of government. By focusing on self-governing and enabling modes, different forms of power play a role, including but not limited to providing financial incentives (Allen 2004). Climate change action, therefore, both requires and facilitates the emergence of 'generative power' – the power to learn new practices and create new capacities (Coaffee and Healey 2003; Bulkeley and Kern 2006).

At the same time, governing by enabling and self-governing both require significant facilitative capacities to co-ordinate action across different policy domains, sectors and societal arenas. Ten years after Bulkeley and Kern's seminal work, researchers continue to develop analytical frameworks for understanding the efficacy of climate change action at the level of the city and local government. Hoppe et al. (2016), in their study of four Dutch cities, suggested adding a fifth type of governing to Bulkeley and Kern's four ideal types: horizontal, networked local climate change governance, in this case across organisational boundaries. Many local governments leading in climate change action have indeed developed such capacities for horizontally collaborative and networked forms of governing. Their role as 'local authorities' is no longer limited to being exclusively focused on their own communities or limited to being 'service providers'. Rather, they are acting as key local-scale facilitators who enable and connect private sector entities, local community groups in their local government area, as well as proactively engaging with neighbouring local governments, higher levels of government and their agencies with a local or regional presence. In this milieu, new and potentially transformative climate change action can emerge that distribute and share the responsibility for reducing greenhouse gas emissions and the local impacts of climate change among a wide group of local stakeholders.

Given the heterogeneity of how climate change action is facilitated through such emerging forms of networked governance, it is useful to examine different forms of governing that can be observed empirically at local and regional spatial scales. The active verb *governing* is given preference over the more static governance here, as we are particularly interested in the proactive and multi-faceted climate change planning and decision-making processes that local actors engage in, as a result of specific motivations and goals. For the reasons outlined above, we are less interested here in the rules and regulations that are created per se but in processes that bring about appropriate conditions for collective action in a given political context (Stoker 2004).

Vertical co-ordination: enabling 'downstream' action

The critical need for multi-level governance in responding to climate change has been highlighted in the literature, including achieving better horizontal and vertical co-ordination between levels of government and departments and across organisations and community groups (Armitage et al. 2010; Hanssen et al. 2013; Bauer 2014). The relationship between different levels of government and sectors will have a bearing on the capacity to co-ordinate and implement climate change responses effectively.

National and regional government provide important 'enabling' contexts for local actors to operate effectively in the development and implementation of strategies and action plans and in supporting capacity building, skill development and knowledge sharing. In the Australian context, climate policy and leadership from higher levels of government has been highly contested and extremely variable over recent decades, leaving local governments and other actors largely leading action, albeit in a highly fragmented way (Moloney and Horne 2015; Moloney and Fünfgeld 2015). Federal-level climate change mitigation policies, such as the infamous 'carbon tax', were introduced at great political cost, only to then be scrapped again by the next incoming government. At state-government level, authorities have been mired in similar political dramas, or they have been slow or reluctant to demonstrate leadership at the state level in the absence of a strong and reliable climate policy at the federal level. It is only in the last few years that some state governments, notably, South Australia, New South Wales and Victoria, have seriously engaged with climate change mitigation and adaptation action at the regional scale.

The role of higher levels of government in either enabling or disabling local action is not always straightforward, as the Australian context has shown. In Australia, a substantial growth of local climate change action has been observed since the early 2000s, despite – or perhaps precisely because of – a constrained national and regional policy context (McGuirk et al. 2014; Moloney and Fünfgeld 2015). That said, there is little doubt that strong political and financial support from higher levels of government is important in catalysing and co-ordinating local action, which has been highlighted in other contexts around the world, for example in Sweden (Granberg and Elander 2007; Nilsson et al. 2012) and Germany (Monstadt 2007; Emelianoff 2013) where national funding and policy frameworks are provided to municipalities to develop climate change strategies and action plans.

Horizontal intra-municipal collaboration

Effective horizontal co-ordination of climate change action is equally important. Climate change mitigation and adaptation implicate a range of activities across government, industry and business sectors, including urban development and planning, transport planning and infrastructure provision, energy and water provision, agriculture, public health, education, finance and procurement. It is therefore vital that a balance is struck between customised sector-specific approaches and cross-sectoral coordination and cooperation, in order to avoid duplication of effort, make the most of synergies, and reduce the risk of maladaptation (Barnett and O'Neill 2010, Juhola et al. 2016).

The diverse challenges of vertical and horizontal co-ordination at national and regional scales are mirrored within a single public sector organisation. Here, those charged with leading climate change action in public sector organisations are frequently situated in marginal roles from where they are working from a "position of relative institutional weakness" (Aylett 2015). Commonly, they work in environmental or sustainability departments – particularly in cases where the organisation is in the early phase of considering climate change actions in their planning and operations and climate change hasn't been firmly established as an organisational agenda. Typically, these individuals are junior or mid-range level officers, burdened with the responsibility of mobilising or co-ordinating climate change initiatives across departments. Without the support of management, these 'bottom up' sustainability led initiatives lack the necessary authority to gain 'buy-in' from other departments who may have little interest or capacity to act. In larger local governments, horizontal coordination can be enabled by in-house integration efforts involving 'joining up' individual organisational units, for example by designating departments with responsibility for co-ordination across service areas and policy units, to ensure that strategic plans and sectoral targets are implemented and monitored (Wilkins 2002; Davies 2009).

One of the ways that municipal governments are encouraging participation across multiple departments in climate change planning and implementation is through intra-municipal networks (Aylett 2015). These 'internal' networks are adopting an enabling mode of governing in that they rely on collaboration, reciprocity and a close working relationship between individuals. A global survey of local governments found that building intra-municipal networks that supported the development of informal, trusted relationships were more effective than those initiatives focused on formal training and education and actions to integrate climate change considerations into budgeting processes or human resources contracts (ibid.). The role of informal networks based on trust becomes particularly interesting when examined in the context of formal efforts – here, trusted relationships may be critical for supporting more authoritative and service-oriented forms of governing.

Horizontal inter-municipal partnerships

Beyond the boundaries of the organisation, climate change action calls for municipalities to actively reach out to other municipal administrations, business owners, civil society groups and individual residents to form effective partnerships for mitigation and adaptation. Here, inter-municipal partnerships can play an important coordinating role, especially with regard to the multi-sectoral task of climate change adaptation. Climate change impacts are rarely contained within municipal jurisdictions but rather extend across multiple boundaries. Single heatwave or extreme rainfall events, but also coastal storm surges, typically affect more than one neighbouring administrative district. Here, the biophysical characteristics of the events call for a coordinated response from the administrations involved. Because adaptation actions in one area can impact positively or negatively on neighbouring areas, coordinating responses across municipalities is critical, not only from an administrative efficiency point of view but also from a reputational and political perspective. Where one local government is seen to be proactive in dealing with sea level rise, for instance, by building a sea wall, this structure may negatively affect the neighbouring area, where the municipality has been more 'conservative' with implementing hard infrastructure measures. The potential for conflict is inherent in unco-ordinated adaptation responses at the local level, including the physical risks to assets and communities as well as political and financial implications.

Inter-municipal co-ordination therefore is essential for adaptation actions and can greatly assist in developing spatially coherent mitigation strategies. Focusing on small to medium-sized cities, Hoppe et al. (2016, p. 23) remark that "inter-municipal networking pays off" and call for further research on inter-municipal network collaboration and the role of citizens in climate change mitigation and on modes of governance that help to empower them.

Trans-municipal networks

In addition to geographically bounded inter-municipal collaboration and co-ordination, regional, national and especially international trans-municipal municipal networks on climate change action have grown rapidly in number, size and relevance. Organisations such as ICLEI, C40 and the Rockefeller Foundation's 100 Resilient Cities campaign have played a significant role in particular in relation to leadership and facilitating knowledge exchange and learning among cities around the world. There is evidence that these global networks are particularly influential in the early stages of policy development, in both climate change mitigation and more recently, also in adaptation (Fünfgeld 2015). In a review of municipal roles and challenges in terms of addressing climate change, a range of factors were identified that shape the nature of these networks' role and contribution: knowledge, leadership, competencies, resources, politics, planning frameworks, jurisdictional 'fit', vertical and horizontal integration between government agencies, networking, partnership and public involvement (Kern and Bulkeley 2009).

Trans-municipal networks and partnerships have also emerged at the subnational scale and operate in a similar way to provide leadership, guidance and learning across their members. Multilateral trans-municipal partnerships across regional and city scale typically include state and non-state actors, private businesses, academic institutions and community groups.

Enabling modes of governing in Australian municipalities

In the following, we examine two specific collaborative climate change action settings at the municipal and inter-municipal scale in the state of Victoria, Australia. One of them highlights intra-organisational collaboration on climate change adaptation at the local administrative scale in the city of Greater Geelong, while the other describes a regional-scale climate change collaborative, where multiple municipalities collaborate with non-governmental actors to make progress on climate change mitigation and adaptation in metropolitan Melbourne.

Australian municipalities provide an interesting case study with regard to the growing role of collaborative and networked forms of local climate change governance. Local governments in Australia are a product of state statute and have limited political and financial capacity funded largely through rates. In the context of a hostile and uncertain political landscape concerning the governing of carbon mitigation in Australia, municipalities have, however, emerged as key actors involved in a broad range of low-carbon interventions incorporating behaviour/social change; retrofitting/demonstration; transition and advocacy (McGuirk et al. 2014). A recent national survey of Australian local government carbon reduction initiatives positions them as "experiments that are exploring institutional partnerships to expand governance capacity and authority" (McGuirk et al. 2015: 50) with the potential to shape urban political and socio-technical transformations.

In terms of climate change adaptation responses, Australian municipal administrations are also playing a leading role, particularly in developing risk and vulnerability assessments, adaptation plans and one-off projects (Measham et al. 2011, Preston et al. 2011). Adaptation measures need to be responsive to particular contexts and needs and encompass all sectors, involving infrastructure provision and retrofitting, flood management and mitigation, disaster response and preparedness, emergency and health services. Local governments play an important role in adaptation due to their connections with, and understanding of, local communities' needs and issues and their role in local service provision, which, conversely, requires integrating adaptation into capital works, asset management and other operational areas. This necessity enlists a wide range of actors across public, private and community sectors and involves a level of institutional complexity beyond that required for mitigation responses.

The two Australian examples presented in the following provide insights into the practical workings of municipal collaboration on climate change action. The first case involves a local government actively working to improve intra-municipal collaboration through developing streamlined decision-making tools across the organisation, and the second case is an example of inter-municipal collaboration involving groups of councils forming 'greenhouse alliances' to mobilise member council responses to climate change and to tackle climate change mitigation and adaptation at a regional scale.

Intra-municipal collaboration for adaptation in the city of Greater Geelong

The city of Greater Geelong is a local government located approximately 75 kilometres south-west of Melbourne in the state of Victoria. With over 191,000 urban inhabitants, Geelong is the second-largest city in the state, while the city of Greater Geelong (CoGG) services a population of over 229,000. Geelong faces major threats from sea level rise and coastal inundation, but also from bushfires – which are projected to increase in frequency and in intensity – flooding and (particularly in its urban areas) from heatwaves.

In 2009, the CoGG decided that the impacts of climate change on the local government area, its people and natural and built assets be assessed and that a climate change adaptation strategy and action plan be developed. To start with, the city took the common path of tendering climate change risk assessment work and working with a consulting firm to evaluate climate change risks and develop an adaptation strategy (City of Greater Geelong 2011). Through careful steering of this process by CoGG and by working with a suitable consultancy with experience in participatory planning, the process included significant involvement from across the local government administration, which helped create co-ownership for adaptation responses and raise awareness across the organisation on the risks and current and future impacts of climate change on the city.

Already at the strategic planning stage, it was decided that a customised process would be needed to develop an in-house planning and decision-making process for dealing with climate change risks effectively, coherently and involving all parts of the organisation – despite the fact that adaptation planning was till then driven by the environment department. Because CoGG already had developed an in-house risk management system, including climate change risks in that system seemed not only logical from a governance and procedural point of view, it also provided an opportunity for improving the risk management system and increasing ownership for in-house risk management broadly, by adding climate change risks.

Designing this in-house climate change risk management process was included as a key action in the climate change adaptation strategy. It was also clear that, in order to develop a proper process, a significant amount of research and customised development would need to be carried out, including understanding existing risk-management processes, assessing their suitability to deal with climate change risks, and developing tailor-made 'decision pathways' for considering climate change risks in all of the organisation's decision-making.

To tackle this significant task, the organisation enlisted a consortium of management consultants and climate change adaptation researchers, who worked together to develop a customised process toolkit for climate change adaptation (Eyre et al. 2012). The toolkit was designed over a 12-month period that included a needs assessment to scope out the purpose and functions of the process, multiple stages of experimenting and testing different process elements with two different departments in the CoGG (engineering services and community development), as well as parallel research to learn from similar organisational adaptation frameworks and toolkits in other contexts. At an early stage, it was identified that the process needed to do two things: (1) be designed to enable specific adaptation actions in response to already identified risks as part of the climate change risk assessment, and (2) ensure that all major projects and initiatives carried out in the organisation were 'screened' for potential links with climate change and risks – either because they exacerbated existing risks or created new risks, or because they adversely impacted greenhouse gas emissions.

Figure 2.1 provides an overview of the scope of the three climate change planning tools that constituted the entire climate change risk management processes at CoGG. Tool 1 provided a process for exploring and assessing the wider risk context, i.e., it guided users through the process of developing a succinct yet precise problem statement that also highlighted known uncertainties. These statements could be used as inputs to reports and briefing notes to management and council, as well as serving as input for the second part of the process. The second tool focused on a process for developing efficient and effective adaptation actions, in response to the risks identified and assessed in detail as part of tool 1. Tool 3 helped screen each identified climate change action for interactions with existing risk. Here, all project ideas or planned actions – whether they were specifically about climate change or not – could be screened for hitherto unidentified potential consequences, including maladaptation. While these processes – especially the one covered by tool 3 – were considered innovative at the time as they addressed organisational and problem complexity, significant innovation was demonstrated in the co-development of the tools. By co-developing the process steps among local government staff, academic experts and consultants, the tools developed were tied to actual organisational planning and decision-making processes. Iterative process development, interspersed with testing workshops, not only helped ensure that the process tools were fit for purpose, it also generated a space for informal learning and exchange among local government colleagues and between local government staff and 'outsiders'.

The toolkit development process therefore provided a welcome opportunity for intra-municipal collaboration that was useful for forging new internal partnerships, learning about other departments' needs and challenges in relation to climate change adaptation, and for significantly enhancing the profile of adaptation in general, across the organisation. In some ways, the *output* of the toolkit development, while highly useful for internal planning processes, was of secondary importance to the internal engagement and collaboration that was achieved through the *process*.

Regional local government alliances for climate change mitigation and adaptation: inter-municipal climate change collaboration

In Victoria there are ten so-called greenhouse partnerships, or as they are officially called, 'climate change alliances' (CCAs), that formed during the 2000s largely by groups of adjacent councils. The CCAs originally emerged as informal

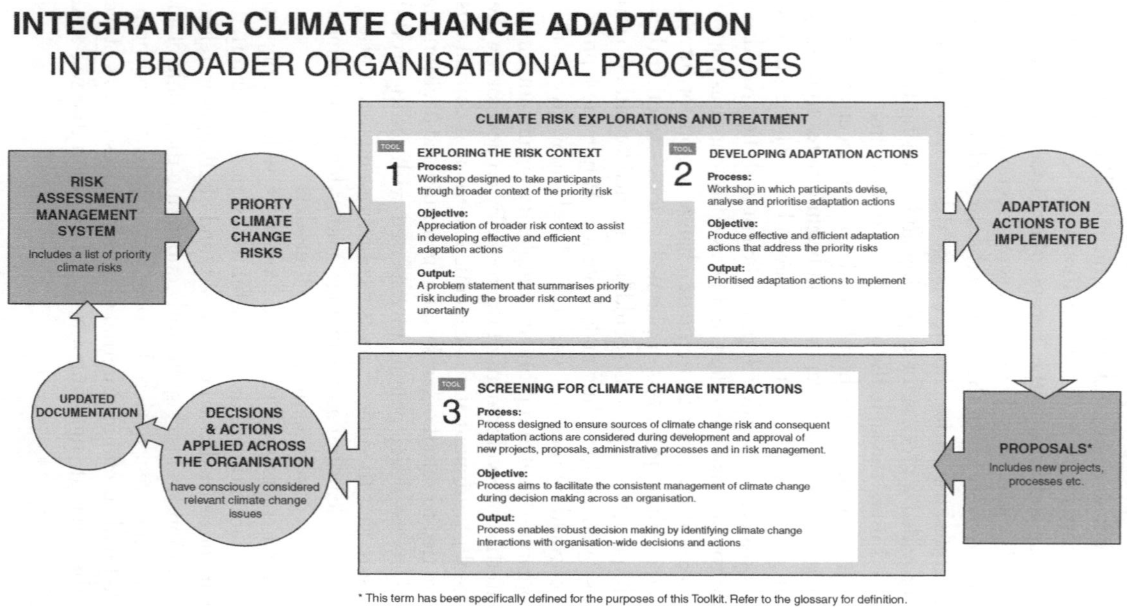

Figure 2.1 Flow chart of city of Greater Geelong process tools as a result of an intra-municipal co-development process involving external stakeholders

The process tools that were developed are numbered from 1 to 3. They focus on better understanding the climate risk context and helping identify specific risk treatment options.

Source: Eyre et al. 2012, p. 7

'information sharing' networks amongst local government sustainability officers driven not only by a shared goal but also by the need to address an emerging policy gap involving local capacity to act on climate change. These networks were facilitated more formally through a state government initiative in 2002, the 'Regional Partnerships Program' as part of the 'Victorian Greenhouse Strategy'. It was recognised that there was a need to build capacity across local governments through strengthening networks and partnerships. The program provided the salary for each alliance to fund an executive officer, whose task was to co-ordinate and mobilise partner councils' mitigation initiatives. This early focus could be characterised as 'self-governing' to improve internal greenhouse actions and capacities, for example through street lighting and energy efficiency programs focusing on council-owned buildings.

The Regional Partnerships Program was reviewed in 2006, and while it was deemed a success in achieving the goals of capacity building through collaboration and project implementation, the state government ceased its annual contributions to alliances in 2008. While there was a period of transition for some of the alliances, most partner councils agreed to provide a small annual fee to maintain the role of the executive officer. These officers have played critical roles in maintaining and building momentum amongst member councils, ensuring the continuity of initiatives, experimenting with different partnerships and projects and progressing long-term regional strategies.

Climate Change Alliances (CCAs) are filling an institutional and policy gap in climate governance in Victoria. Through processes of collaboration, they are enabling local governments to overcome, to some extent, their political and financial weaknesses and confront the short- and long-term challenges local government faces in dealing with climate change. The processes of collaboration involve mediating both horizontally across local government departments and boundaries and across different institutional practices, as well as vertically between groups of councils (regionally) and state and federal governments and, increasingly, with civil society, businesses and NGOs (Moloney and Horne 2015; Moloney and Fünfgeld 2015).

We focus briefly on one climate change alliance, the Western Alliance for Greenhouse Action (WAGA), which has been operating for over ten years and has eight member councils covering Melbourne's western region and the city of Geelong. Among its key initiatives, two are noteworthy and demonstrate different forms of collaboration and leadership through 'enabling' innovation at a regional scale. The first is its 'Low Carbon West Strategy (LCWS)' (WAGA et al. 2014) and the second is its 'How Well Are We Adapting?' project (WAGA 2014–2017), which focuses on developing a framework for councils to monitor, evaluate and report on progress on climate change adaptation.

The LCWS was developed by WAGA with project partners LeadWest and Regional Development Australia (RDA) (Western Melbourne) and involved a number of other private industries and businesses from the region. LeadWest (which is constituted as a company limited by guarantee) includes six local governments (also members of WAGA) and involves major companies and other organisations, each with substantial operations and interests based in Melbourne's west. The purpose of collaborating with agencies with diverse interests was to enable the western region of Melbourne to work towards the shared goal of a 'low carbon economy'. The strategy "seeks to position the West as a centre for the low carbon economy in Australia" (WAGA Exec Officer, Aug 2015), and aims to create opportunities to 'prototype' and 'market test' actions with industry, council and community through an iterative process of experimentation, evaluation and learning. The strategy is referred to as 'a transitional strategy for the region' to support the region's growth while limiting the increase in GHG emissions (WAGA et al. 2014).

The partners enlisted two large consultancies to develop the strategy and lead the consultation process and analysis. After two years researching and developing Low Carbon West, a set of twenty actions were developed across four sectors (business and industry, communities, transporting people and freight, urban growth and development). Each identifies the relevant agencies and actors needed to mobilise actions from within the network to beyond including state government agencies, civil society groups and the private sector. While the strategy-development process was the outcome of a successful collaboration between actors with a diverse range of interests, the current phase of mobilising actions requires more than 'governing by enabling'. Other forms of governing will be required to enlist private sector actors, for example through regulatory means or other forms of incentive (i.e., governing by authority) and as such will require a stronger role from state and federal governments.

The second initiative, 'How Well Are We Adapting?' (WAGA 2014–2017), involves a collaboration led by three WAGA member councils to develop a framework to monitor and evaluate (M & E) climate change adaptation across the WAGA region. Supported by grant funding from the Victorian state government, the aim was to create, through a participatory process, a 'fit-for-purpose' model for M & E, which could be used not only by member councils but potentially by all local governments in Victoria. WAGA invited a number of other local governments from metropolitan Melbourne, including other alliances, to engage with and learn from the project as 'observer' councils. WAGA recognised that they were generating new knowledge to assist local governments in their adaptation work. However, they were also focused on ensuring that the framework for M & E was responsive to their member councils' needs. With a focus on key council service areas implicated in climate change adaptation, the project involved collaborating horizontally across member councils and across different departments within councils. This was challenging as not only does each council have different organisational structures and cultures, but service areas, including emergency services, open space and water management, assets and infrastructure, and planning and development, are not necessarily engaged with climate change adaptation nor the need to M & E their progress on outcomes in this area. The collaborative process therefore focused around building relationships with different officers and managers in different councils' departments, taking time to learn about their key issues and priorities, engaging in knowledge sharing events around climate change adaptation, and building the capacity of different departments to shape and engage with a M & E process. The primary goal in this collaborative process could be characterised as promoting mutual understanding and consensus around the development and implementation of an M & E framework for CCAs. WAGA partnered with a university and environmental consultant to develop both the framework and the consultation process for the project. The process of piloting the framework in three councils is currently under way. At the time of writing, the lessons from this project are being drawn upon in the current work of state government, as it develops a monitoring and evaluation framework for climate change adaptation at the state level.

Conclusion

Returning to the four modes of governing identified by Bulkeley and Kern (2006), one of the key conclusions from their discussion around forms of collaborative governance and our brief analysis of two case studies in Australia was to highlight innovations in local governance and also identify some of the challenges for local governments in terms of their capacity to act through, and co-ordinate, different modes of governance. In more recent research on the role of local governments responding to climate change, Aylett (2015) highlighted some of the ways in which local governments are endeavouring to build their governing capacity, particularly through networks and partnerships, and that we need to better understand how these work in practice. This experience of emerging networks at the scale of local climate change action is consistent with the broader recognition that forms of collaborative governance are necessary to deal with the problems and institutional complexity inherent in mobilising climate mitigation and adaptation efforts. In this chapter, we have identified some of the characteristics of collaboration and explored how they are emerging in two particular cases in Victoria, Australia, through intra-municipal and inter-municipal networks.

In the city of Greater Geelong case, intra-municipal collaboration was both an enabler for more coherent climate change adaptation planning and a necessary ingredient for translating a high-level strategy into more concrete actions, by integrating climate change risks into the organisation's risk-management system. Collaboration was also key in developing an organisational process for considering risks across the organisation and for developing a concrete, fine-grained plan for how climate change risks could best be included as part of routine organisational risk management. The fact that in-house teams from different departments were involved in developing and testing the adaptation planning and decision-making process, as mapped out in the Climate Change Adaptation Toolkit, increased co-ownership of the toolkit and resulted in general awareness-raising on the need to consider climate change impact and invest time and effort in adaptation.

In the case of the climate change alliances (CCAs), the very existence of these alliances, which is unique to the state of Victoria, is due to the recognition that building local capacities to respond to climate change could only be achieved through networks and partnerships. This process of regional partnership formation emerged largely through informal processes of municipal sustainability officer networks and later through some state government funding. Their ongoing work, however, is currently sustained by annual voluntary contributions from member councils, who continue to see the value in regional collaboration. As local governments are politically and financially weak in Australia, regional collaborations are opening up new avenues for strategic planning responses to climate change in a context of uncertainty at both state and federal levels. The CCAs can be understood as enabling processes of 'self-governing' across their regions (see Bulkeley and Kern 2006) in that they are interested in both responding to and shaping the priorities of their member councils in acting on climate change by developing regional strategies and assessments, trialling new projects and processes, and building internal capacities. They are focused on 'governing by enabling' through their networks. As alliances they are enabling their member councils to integrate climate change across their organisations and to strengthen the capacity of councils to enable their communities to better respond to climate change. They do this through creating new forms of knowledge and knowledge management systems (i.e., M & E framework for CCAs) and through experimenting with new initiatives enlisting regional development agencies and private sector actors (i.e., Low Carbon West Strategy). While CCAs demonstrate many characteristics of collaborative governance, there are a number of challenges in terms of ensuring ongoing implementation and effective action at the local government scale. CCAs do not have the authority and autonomy to act in many instances around implementing particular goals, for example around achieving energy efficiency goals in the building and industry sectors as this involves other actors and modes of governing (i.e., state and federal regulators). They can, however, continue to develop their networks to include relevant actors with shared goals. As they continue to pursue strategic goals and innovate around different projects, state government and other actors, who may traditionally 'lag' in their response, will be able to learn from their experience. This is already the case as the state government is learning from WAGA in its current work around developing an M & E framework for the state. While local governments are to some extent weak in terms of their capacity to 'govern by authority' or 'by provision' in terms of responding to climate change, there is little doubt about the role of networks and alliances in building their capacity to govern by enabling.

References

Allen, J. (2004). The whereabouts of power: Politics, government and space. *Geografiska Annaler, Series B: Human Geography*, 86(1), 19–32.

Allman, L., Fleming, P., and Wallace, A. (2004). The progress of English and Welsh local authorities in addressing climate change. *Local Environment*, 9(3), 271–283.

Armitage, D., Berkes, F., and Doubleday, N. (2010). *Adaptive Co-Management: Collaboration, Learning, and Multi-Level Governance*. British Columbia: UBC Press.

Aylett, A. (2015). Institutionalizing the urban governance of climate change adaptation: Results of an international survey. *Urban Climate*, 14, 4–16.

Barnett, J., and O'Neill, S. (2010). Maladaptation. *Global Environmental Change*, 20(2), 211–213.

Bauer, A. (2014). Multi-level governance of climate change adaptation through regional partnerships in Canada and England. *Geoforum*, 51, 121–129.
Bidwell, R.D., and Ryan, C.M. (2006). Collaborative partnership design: The implications of organizational affiliation for watershed partnerships. *Society & Natural Resources*, March, 19, 827–843.
Bulkeley, H., and Kern, K. (2006). Local government and the governing of climate change in Germany and the UK. *Urban Studies*, 43(12), 2237–2259.
City of Greater Geelong, (2011). *Climate Change Adaptation Strategy*. Geelong: City of Greater Geelong.
Coaffee, J., and Healey, P. (2003). 'My voice: My place': Tracking transformations in Urban governance. *Urban Studies*, 40(10), 1979–1999.
Conley, A., and Moote, M.A. (2003). Evaluating collaborative natural resource management. *Society & Natural Resources*, 16(5), 371–386.
Davies, A.L., and White, R.M. (2012). Collaboration in natural resource governance: Reconciling stakeholder expectations in deer management in Scotland. *Journal of Environmental Management*, 112, 160–169.
Davies, J.S. (2009). The limits of joined-up government: Towards a political analysis. *Public Administration*, 87(1), 80–96.
Demerit, D., and Langdon, D. (2004). The UK climate change program and communication with local authorities. *Global Environmental Change*, 14, 325–336.
Emelianoff, C. (2013). Local energy transition and multilevel climate governance: The contrasted experiences of two pioneer cities (Hanover, Germany, and Vaxjo, Sweden). *Urban Studies*, 51, 1378–1393.
Eyre, S., Niall, S., Silke, F., and Young, S. (2012). *Climate Change Adaptation Toolkit User Guide: A Comprehensive Guide to Planning for Climate Change Adaptation in Three Steps*. Geelong: City of Greater Geelong, NetBalance Foundation and RMIT University.
Fünfgeld, H. (2015). Facilitating local climate change adaptation through transnational municipal networks. *Current Opinion in Environmental Sustainability*, 12, 67–73.
Gash, A. (2016). Collaborative governance. In: Ansell, C.K. and Torfing, J. (eds.) *Handbook on Theories of Governance*. Cheltenham: Edward Elgar Publishing, 454–467.
Gerlak, A.K., and Heikkila, T. (2006). Comparing collaborative mechanisms in large-scale ecosystem governance. *Natural Resources*, 46(3/4), 657–707.
Granberg, M., and Elander, I. (2007). Local governance and climate change: Reflections on the Swedish experience. *Local Environment*, 12(5), 537–548.
Hanssen, G.S., Mydske, P.K., and Dahle, E. (2013). Multi-level coordination of climate change adaptation: By national hierarchical steering or by regional network governance? *Local Environment*, 18(8), 869–887.
Hoppe, T., van der Vegt, A., and Stegmaier, P. (2016). Presenting a framework to analyze local climate policy and action in small and medium-sized cities. *Sustainability*, 8(9), 847.
Johnston, E.W., Hicks, D., Nan, N., and Auer, J.C. (2011). Managing the inclusion process in collaborative governance. *Journal of Public Administration Research and Theory*, 21(4), 699–721.
Juhola, S., Glaas, E., Linnér, B-O., and Neset, T.-S. (2016). Redefining maladaptation. *Environmental Science & Policy*, 55, 135–140.
Kern, K., and Bulkeley, H. (2009). Cities, europeanization and multi-level governance: Governing climate change through transnational municipal networks. *Journal of Common Market Studies*, 47(2), 309–332.
Kinnear, S., Patison, K., and Mann, J. (2013). *Network Governance and Climate Change Adaptation: Collaborative Responses to the Queensland Floods*. Gold Coast: National Climate Change Adaptation Research Facility.

Koontz, T.M., and Thomas, C.W. (2006). What do we know and need to know about the environmental outcomes of collaborative management? *Public Administration Review*, 66(1), 111–121.

Lasker, R.D., and Weiss, E.S. (2003). Broadening participation in community problem solving: A multidisciplinary model to support collaborative practice and research. *Journal of Urban Health: Bulletin of the New York Academy of Medicine*, 80(1), 14–60.

Linnenluecke, M.K., and Griffiths, A. (2013). Firms and sustainability: Mapping the intellectual origins and structure of the corporate sustainability field. *Global Environmental Change*, 23(1), 382–391.

McGuirk, P., Dowling, R., and Bulkeley, H. (2014). Repositioning urban governments? Energy efficiency and Australia's changing climate and energy governance regimes. *Urban Studies*, 51(13), 2717–2734.

McGuirk, P., Dowling, R., Brennan, C., and Bulkeley, H. (2015). Urban carbon governance experiments: The role of Australian local governments. *Geographical Research*, 35(1), 39–52.

Measham, T.G., Preston, B.L., Smith, T.F., Brooke, C., Gorddard, R., Withycombe, G., and Morrison, C. (2011). Adapting to climate change through local municipal planning: Barriers and challenges. *Mitigation and Adaptation Strategies for Global Change*, 16(8), 1–21.

Moloney, S., and Fünfgeld, H. (2015). Emergent processes of adaptive capacity building: Local government climate change alliances and networks in Melbourne. *Urban Climate*, 14, 30–40.

Moloney, S., and Horne, R. (2013). Low carbon transitions: A melbourne case study. In: Ruming, C., Randolph, B., and Gurran, N. (eds.) *Proceedings of the State of Australian Cities Conference, 26–29 Nov 2013, Sydney, Australia*. Sydney: State of Australian Cities Conference.

Monstadt, J. (2007). Urban governance and the transition of energy systems: Institutional change and shifting energy and climate policies in Berlin. *International Journal of Urban and Regional Research*, 31(2), 326–343.

Newman, J., Barnes, M., Sullivan, H., and Knops, A. (2004). Public participation and collaborative governance. *Journal of Social Policy*, 33(2), 203–223.

Nilsson, A.E., Gerger Swartling, Å., and Eckerberg, K. (2012). Knowledge for local climate change adaptation in Sweden: Challenges of multilevel governance. *Local Environment*, 17(6–7), 751–767.

O'Brien, K., Reams, J., Caspari, A., Dugmore, A., Faghihimani, M., Fazey, I., . . . Winiwarter, V. (2013). You say you want a revolution? Transforming education and capacity building in response to global change. *Environmental Science & Policy*, 28, 48–59.

Preston, B.L., Westaway, R.M., and Yuen, E.J. (2011). Climate adaptation planning in practice: An evaluation of adaptation plans from three developed nations. *Mitigation and Adaptation Strategies for Global Change*, 16(4), 407–438.

Stoker, G. (2004). *Transforming Local Government: From Thatcherism to New Labour*. Basingstoke: Palgrave Macmillan.

Taylor, B., and de Loë, R.C. (2012). Conceptualizations of local knowledge in collaborative environmental governance. *Geoforum*, 43(6), 1207–1217.

Western Alliance for Greenhouse Action (WAGA), LeadWest, Regional Development (Western Melbourne) (2014). *Low Carbon West: A Strategy for the Transition to a Low Carbon Economy in the WAGA Region*. Retrieved from http://waga.com.au/climate-change-action/low-carbon-west/

Western Alliance for Greenhouse Action (WAGA) (2014–2017). '*How Well Are We Adapting? A Framework to Monitor, Evaluate and Report on Climate Change Adaptation in Western Melbourne, Project*. Retrieved from http://waga.com.au/climate-change-action/how-well-are-we-adapting/

Wilkins, P. (2002). Accountability and Joined-up government. *Australian Journal of Public Administration*, 61(1), 114–119.

3 Governance and agency beyond boundaries

Climate resilience in Port Vila's peri-urban settlements

Alexei Trundle

Introduction

Vanuatu's capital Port Vila lies on the southwestern coast of Efate, the central island of the Melanesian archipelago. Situated on the Pacific 'Ring of Fire' and to the east of the New Hebrides Trench, the country is a geological product of seismic uplift, tectonic movement and volcanism – processes that its inhabitants continue to face today in the form of tsunamis, earthquakes and eruptions. However, it is climate-related extreme events that most contribute to Port Vila's high level of natural hazard vulnerability (The World Bank 2009). In a recent global risk assessment, Vanuatu was ranked as the most disaster-exposed nation out of 171 countries for the fifth year running (Mucke 2015). Port Vila itself was identified in a report by global risk consultancy Verisk Maplecroft as the most natural disaster exposed of a global sample of 1300 cities – findings published only three days before the city was devastated by the impacts of Tropical Cyclone Pam in March 2015 (Verisk Maplecroft 2015).

Relatively stable year-round temperatures mask an extremely variable inter-annual distribution of rainfall. Driven by both its location within the South Pacific Convergence Zone and the cyclical movement of the El Niño Southern Oscillation across the region, Port Vila's annual rainfall record of more than 4000mm is almost twice its long-term average, and five times the recorded minimum; 800mm in 1947 (PACCSAP 2014). An average of 2.4 tropical cyclones pass through the Vanuatu Exclusive Economic Zone per year, but again this number varies greatly; six occurred in the 1991–92 cyclone season, compared with none during 2002–03 (ibid). With most of the country's inhabitants reliant on rivers, springs and unconfined aquifers for drinking water (as well as crop production, livestock and livelihood products), El Niño–driven drought puts significant stress on local economies, food production and human health (Trundle and McEvoy 2015). Land- and marine-based extreme heat events, although less well documented, have also been observed to lead to coral bleaching, freshwater 'fish kills' and human health issues (Johnson et al. 2015).

Historical exposure to these climatic variations and extremes has resulted in high levels of community-based coping capacity and a wealth of traditional knowledge, technologies and management techniques, many of which continue to be applied by the *ni*-Vanuatu[1] population today (Ensor 2015). However, in settings such as Port Vila traditional livelihoods, knowledges and *kastom* (customary) social structures operate in isolation from their geographic origins, alongside the post-colonial urban institutions and laws that form the basis of municipal governance. In peri-urban areas the conflict between these two systems is particularly acute. Migrant inhabitants of these zones reside beyond the remit of municipal governance structures and services, but at the same time are limited in their access to customary social systems, resources and knowledge.

This chapter will examine the dualities of urban governance and agency that occur in Greater Port Vila's peri-urban spaces, and assess them in light of efforts to enhance the city's climate resilience. First, a historical overview of Port Vila is provided to contextualise the meaning of 'urban', and its interface with more traditional forms of governance in a post-colonial Vanuatu. The continuity of *kastom* and traditional knowledge in the identities and local social systems of peri-urban dwellers will then be reflected on, in comparison with institutional and scientific constructs of both local climatic baseline conditions and projected and observed climate change. The interplay between these parallel knowledge and governance systems in Port Vila's peri-urban growth areas is then used to highlight a number of ways that local environmental knowledge, agency and adaptive capacities could be integrated into Port Vila's urban development and climate adaptation planning.

Rather than perpetuating the structural dualities that have led to this capacity deficit, it is proposed that these local modes of knowledge, historically excluded from city decision-making processes, provide a crucial source of community-level resilience, particularly during extreme climate events. Drawing on empirical evidence from first-hand accounts of Tropical Cyclone Pam and the city-wide 'Planning for Climate Change' project led by the United Nations Human Settlements Programme (UN-Habitat) and RMIT University, opportunities for building climate resilience from the local scale upward – in collaboration with expert scientific advice – demonstrate an underutilised avenue for engaging climate change action. Conversely, enactments of transformative change that excludes these informal modes of climate resilience can be demonstrated to lead to perverse adaptation outcomes, undermining existing adaptive capacities in informal spaces that have the potential to complement – and in some cases, exceed – those held by formal governance institutions. These reflections, although specific to Port Vila, have potential application elsewhere, particularly in cities found in other Small Island Developing States (SIDS), as well as for climate planning in informal settlements more generally, which house an estimated 850 million urban inhabitants globally (UN-Habitat 2012).

'*man Vila*': resisting an urban identity

In order to understand current-day governance and agency in Port Vila's peri-urban domain, the city's relatively short history provides some useful context; the Vanuatu capital having been central to the country's colonial occupation, independence movement and the post-colonial consolidation of the nascent nation-state. As with all climate change action planning, characterising the underlying non-climate system (in this case, that of decision-making, governance and agency in the greater Port Vila area) is critical if climate-related interventions are to be effective, relevant, equitable and sustainable (Bulkeley and Tuts 2013). For rapidly growing and developing cities such as Port Vila, non-climate futures can exhibit more uncertainty than projected climatic futures, requiring a similarly sophisticated consideration of urban baseline conditions, recent trends and the historical settings from which the city's current conditions emerged. It is the latter that provides the most compelling causal linkages for the issues that characterise Port Vila's contemporary urban governance, services and infrastructure.

As noted by Rawlings in his historical analysis of urbanisation in Vanuatu, "*it is rare for the term* man Vila *(person from Port Vila) to be used by either rural-to-urban migrants in town or by peri-urban villagers, except perhaps by way of insult*" (1999, p. 76). This widespread resistance to an urban *ni*-Vanuatu identity is juxtaposed by an emphasis on maintaining connections with one's island or rural origins, and underpins the complex governance arrangements that prevail in Port Vila's peri-urban fringe. Such 'anti-urbanism' can be at least partially explained by three features of the city's relatively short history. First, the colonial creation, control and ownership of the city worked actively to spatially exclude customary law and land tenure, as well as restrict the physical presence of *ni*-Vanuatu in the urban domain (Woodward 2014). Second, the subsequent disbandment of colonially resourced urban processes and planning resulted in a degradation and decline of municipal services, infrastructure and utilities. Third, an inability to reconcile pre-colonial *kastom* land rights in the post-independence era has resulted in politically reinforced but unrepresentative municipal boundaries, leaving peri-urban governance to remain disjointed from Port Vila's established urban core (Chung and Hill 2002). Despite this, more *ni*-Vanuatu continue to move to the city in search of employment, services and alternative livelihoods. Spatially, it is the city's peri-urban areas that are growing most quickly, yet these same areas face a vacuum of formal urban governance and a physical and institutional disconnect from services and infrastructure (Trundle and McEvoy 2015).

Port Vila's municipal boundaries were first delineated in 1929, the colonial agricultural township having grown to a population of around 1000 from an estimated 200 in 1906 (Rawlings 1999, p. 83). Although much of the area was purchased for use as a plantation by Frenchman Ferdinand Chevillard in the late 19th century (Lightner and Naupa 2005), the nature of the acquisition from Port Vila's customary owners (communities from the surrounding villages of Pango, Erakor and Ifira), and the area's agreed spatial extent, have subsequently been contested and were poorly recorded (Woodward 2014). Notably, these transactions took place during a complex political re-arrangement of colonial administration, a rapid uptake of Christianity (driven by French and English missionary groups) and widespread depopulation across the archipelago driven by a series of devastating epidemics, relatedly brought about by foreign-borne diseases (Rawlings 1999).

The legislated repatriation of all 'alienated' land following the country's independence in 1980 proved to be particularly problematic in Vanuatu's two urban centres: Port Vila and Luganville. Following an unsuccessful attempt to provide customary owners with regular land rent payments in the capital – culminating in riots in 1988 – compensation payments for the permanent loss of land tenure were successfully made to two of the three claimant land owner groups in 1992 (Rawlings 1999, p. 76). However, a majority of the third land-owning group from Ifira refused the payment and have never had their claims fully resolved. Although these payments diffused much of the community unrest, tensions between members of all three customary landholder groups and *man aelan* (urban migrants from the other islands) continue to lead to minor scuffles and late-night violence, particularly amongst the city's youth (ibid.).

In large part because of this tumultuous history, Port Vila's boundary has remained almost unchanged in the three and a half decades since independence, other than a minor revision of ward delineations in 2013 (GoV 1980; GoV 2013). This is despite the urban area's population growing at an average rate of more than 4 percent annually over this period, well beyond the municipal limits into the surrounding Shefa Province (GoV 1987, p. 122). An outline of the city – including the formal municipal boundaries, 2009 household locations, and the cadastral boundaries of more recent subdivisions (a proxy for subsequent urban growth) – is provided for reference in Figure 3.1.

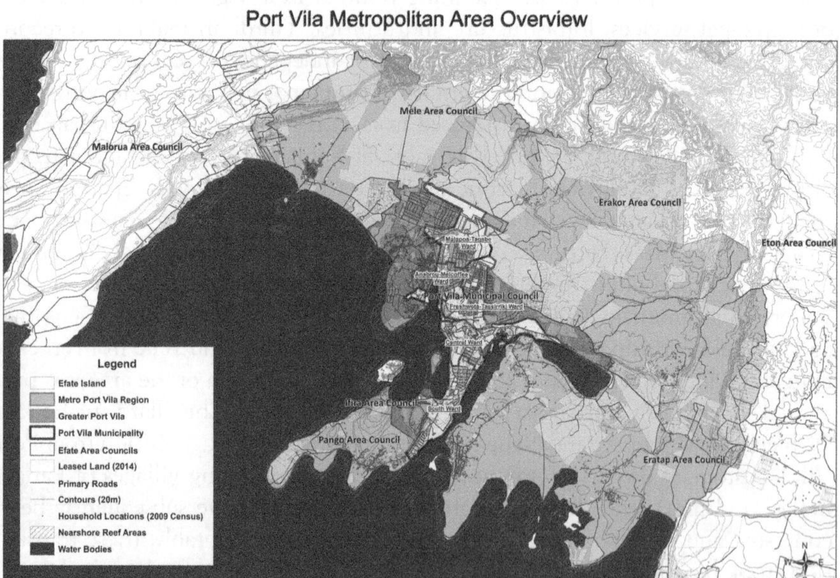

Figure 3.1 Greater Port Vila – household locations (2009), subdivisions (2014) and municipal ward boundaries (2016). Data sourced from the Secretariat of the Pacific Community PopGIS Portal and the Government of Vanuatu

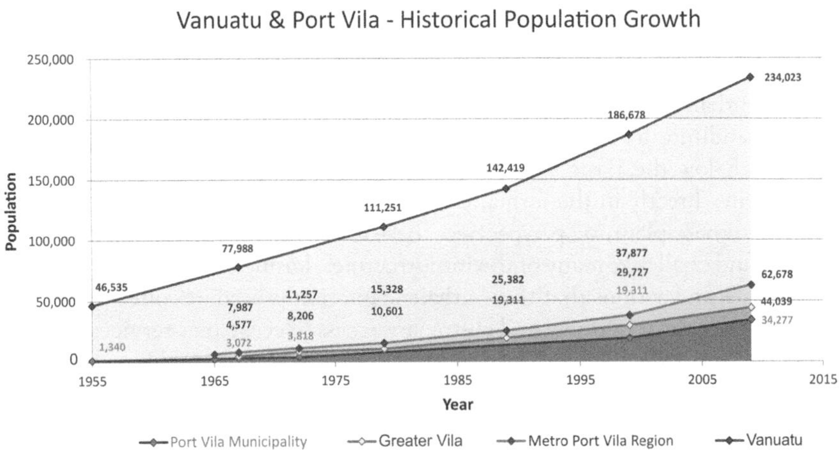

Figure 3.2 Analysis of historical rates of urban and non-urban growth in Vanuatu. Data sourced from GoV 1986; GoV 1989 & VNSO 2010.

Statistics listed in the 2009 national census and numerous other planning documents and policies record Port Vila as having an enumerated population of 44,039 (VNSO 2010, p. 12). However, the urban-classified area included in this count is neither reflective of the city's municipal boundary, nor inclusive of the peri-urban zones and villages in which many urban workers reside. A more detailed analysis, based on the 2009 household counts depicted spatially in Figure 3.1, finds 34,277 citizens living within the municipal boundary and 62,678 residing within the greater Metropolitan Port Vila Region.[2] The population of this wider area over the decade preceding 2009 grew at an average annual rate of 6.55 percent, as shown in Figure 3.2. Excluding this area, the rest of Vanuatu only grew at an average rate of 1.51 percent, demonstrating an accelerating rural-to-urban migratory trend as *ni*-Vanuatu move from rural areas and island townships to capital. From an estimated 2.9 percent in 1955, the metropolitan Port Vila region is now contains more than 26.9 percent of the country's total population.

'*ples*': land rights as livelihoods

Although maps such as that shown in Figure 3.1 are useful reference points for urban planners, researchers and policy makers, for most *ni*-Vanuatu these depictions are not only largely inaccessible but also fail to reflect the communal and non-delineable nature of *kastom* land rights and ownership (McDonnell 2013). For the '*man aelans*' – rural-to-urban migrants originating from other Vanuatu islands – who have settled within the city's peri-urban zone, it is the wider connectivity with subsistence resources that is of equal, if not greater, relevance than the cadastral delineation of housing allotments. The lack of *ples* – literally 'place', but also a core component of land and resource rights and ownership – exhibited by migrants prevents access to ecosystem-based resources such as firewood for cooking fuel, as well as bush gardens and fisheries needed for food and other livelihood supplies. Purchasing goods and services can be even less feasible for rural-to-urban migrants who lack cash incomes. A survey of household income and expenditure in 2010 found that roughly one in seven households in Port Vila are below the Basic Needs Poverty Line, lacking the capacity to procure these items directly in the formal economy (Trundle and McEvoy 2015, p. 11). From a climate-planning perspective, these characteristics are atypical of urban centres and challenge many of the infrastructure-, business-, and housing-focused techniques and actions that these urban adaptation procedures often emphasise. Instead, the role of informal and customary access to ecosystem-services – such as areas for firewood collection, fishing and bush gardening – could be argued to be more fundamental to the livelihoods of a majority of the peri-urban population. By way of illustration, the peri-urban household dependency on crop-based cash income is clearly visible when depicted relative to other areas across Greater Port Vila (see Figure 3.3 below).

The subdivision of land for housing – with varying degrees of legal and customary process – does occur regularly on the customary land outside of the municipal boundary, as can be seen in the lighter-shaded areas shown in Figure 3.1. However, prices for even small subdivisions are well beyond levels affordable for most

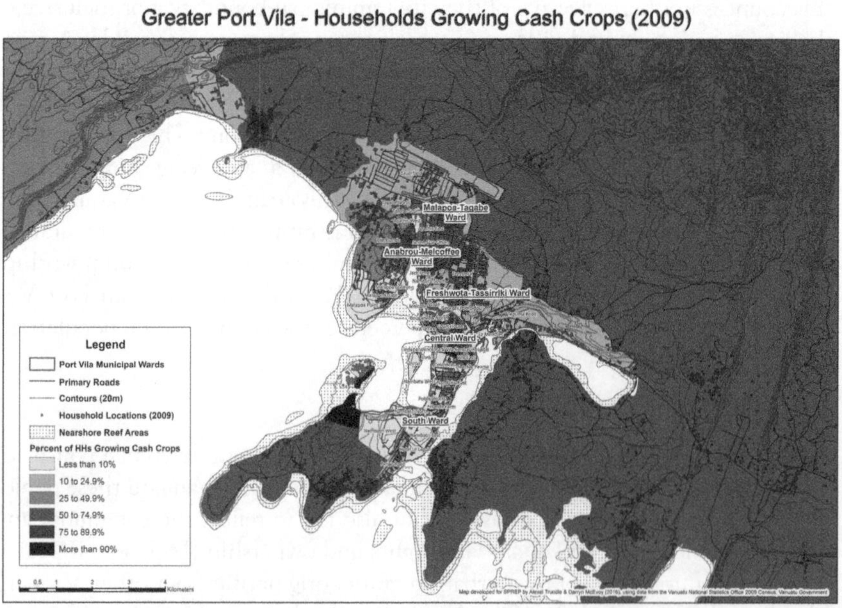

Figure 3.3 Households in the Greater Port Vila area growing crops for cash income, based on 2009 census data (modified from McEvoy et al. 2016)

rural-to-urban migrants, particularly for those households that move to Port Vila transitioning from a customary, subsistence-based livelihood where cash income is generally only required for transport, health and educational services (Banes and Wilson 2011). As a result, extensive informal settlement arrangements have developed over time on a largely ad hoc basis through negotiations with customary land owners, leaseholders and government representatives, in some instances without the knowledge of – or in conflict with – recognised title. This has resulted in a complex array of informal settlement typologies, which operate outside of both customary and state-based understandings of urban governance and planning.

Although analysis by Chung and Hill (2002) went some way to characterising Port Vila's informal communities, limited comprehensive or longitudinal analysis has been undertaken (Mecartney 2000; Leslie 2012). The most recent effort to map the locations of Port Vila's informal settlement areas in 1997 by the Asian Development Bank is also out of date, following the encroachment of new subdivisions, settler evictions and informal population growth driven by in-migration (Chung and Hill 2002, p. 9). A review in 2012 by the National Housing Corporation of 14 of the city's main informal settlements was found to be broadly consistent with groundtruthing prior to the impact of Tropical Cyclone Pam in March 2015 (NHC 2012; Trundle and McEvoy 2015). Analysis of these figures shows that at least a fifth of the city's residents were living under informal tenure arrangements in 2012, with most of these informal dwellers located in peri-urban areas where 39 percent of residents lack legal rights to their housing (NHC 2012).

Developing an understanding of the intersection of the landed and the landless, as well as state-based and customary land governance, is thus a pre-requisite for understanding how governance and agency operate in Port Vila's peri-urban domains. By focusing on the community-level actors who inhabit these informal peri-urban spaces, the impacts of climate change can be framed through the interaction between these actors and the prospective adaptation of their livelihoods, which provide the everyday basis for decision-making at the household scale. In this sense, this chapter focuses on how climate change action can be undertaken in areas where agency is enacted by actors and informal institutions whose actions are not necessarily aligned with the formal domain of state- and city-based governance.

Climate knowledge(s) in peri-urban Vila

Annual average temperature records in Port Vila only date back to 1948, and although rainfall records date back to 1907, significant data gaps exist in both datasets, with sub-annual and daily data being even more limited (BoM and CSIRO 2014, p. 321). Modelling of the spatial distribution and severity of climate hazards across the city is also deficient, with data for the development of flooding, coastal inundation and landslide maps being unavailable at an appropriate scale to assess localities within the city until very recently (The World Bank 2009, p. 16; Heron et al. 2016). Recently released LiDAR-derived hazard mapping by The

World Bank, Geoscience Australia and the Vanuatu National Disaster Management Office is expected to improve these capacities. However, earlier assessments by SOPAC and others mainly focused on geo-hazard risks such as earthquakes and tsunamis (see, for instance, Shorten et al. 2004). Local spatial variations in impacts associated with other extremes such as cyclonic wind and heat exposure remain un-assessed in the Port Vila area.

These limitations in and of themselves warrant consideration of alternative sources of local climate information. However, the process of grounding these scientific forms of knowledge in community-level experiences of climate variability and change, as well as local livelihood dependencies, is argued to be better understood as an opportunity to create unique synergies between local and scientific forms of knowledge for mutual benefit. By integrating 'top-down' climate planning with 'bottom-up' knowledge, dependencies on specific climate variables can be elucidated, adaptation needs prioritised and prospective platforms for resilience-building actions identified. Further, the integration of these socially and scientifically derived forms of climate knowledge can challenge assumed norms within both knowledge-holding communities; those of the lived experience locally, and those embedded within the processes and viewpoints of governance institutions and associated parties. This results in more effective climate adaptation through more accurate prioritisation of climate hazards and risk treatments, as well as a more comprehensive understanding of the knock-on impacts of transformative change at a local scale within urban systems.

The examples below are based on a project that followed such an integrative process, as part of UN-Habitat's Cities and Climate Change Initiative. Led by RMIT University, the project's first phase culminated in development of a Climate Change Vulnerability Assessment for the city, researched over a two-year period prior to 2015 and completed shortly before the impact of Tropical Cyclone Pam in March of that year (Trundle and McEvoy 2015). The research was conducted in partnership with Port Vila Municipal Council (PVMC), Shefa Provincial Government (SPG) and the Vanuatu National Advisory Board for Climate Change and Disaster Risk Reduction (NAB) and made possible through the invaluable contributions of numerous community-based and non-government organisations. UN-Habitat is in the process of developing an action plan based on the findings of this research in partnership with local communities, stakeholders and government, which is due to be finalised in mid-2017.

'Oli no putum name blo climate change'

Port Vila's largest informal settlement area, Blacksands, is located within the city's peri-urban zone, extending westward from the municipal boundary of the Malapoa-Tagabe Ward along the banks of the Tagabe River to its mouth at Mele Bay (labelled on the map in Figure 3.3). The area has been inhabited on an informal basis since the early 1960s under an agreement with the area's customary owners, the Ifira Land Council (Hardwick 2012). By 2009 the settlement had grown to a population of over 7,000 (NHC 2012). Based on an empirical assessment of geographic features alone, the settlement is one of Port Vila's most highly exposed to climate-related hazards. Blacksands occupies a low-lying coastal floodplain downstream of the urban area, with no shelter from prevailing winds, and no formal access to the city's water supply (Trundle and McEvoy 2015). However, despite many of Blacksands' residents being dependent on local ecosystem services rather than cash incomes for their livelihoods, scientific data on the Tagabe River's downstream water quality, saline ingress into the local water table and longitudinal changes to coastal morphology and marine resources are not available (Trundle and McEvoy 2015; Tawney 2006, p. 71).

Despite the aforementioned limitations of scientific data, the undisrupted presence of these informal communities for over 50 years (structured around island origins and led by a core group of established families) provided an alternative, bottom-up avenue for developing detailed reconstructions of the impacts of past extreme events and observations of recent trends. For instance, although coastal erosion in the area is not documented in any scientific or peer-reviewed publications, community elders were able to visibly demonstrate the extent of the erosion of the settlement's distinctive black volcanic sand during Cyclone Uma in 1987, as well as infill of coral sands in the coastal zone due to subsequent storm surge events (see Figure 3.4).

This information was gathered through workshops and transect walks with members of five of the Blacksands communities in 2013. During these activities, it was noted by participants that although the area had high levels of current climate exposure, socio-economic disadvantage and infrastructure sensitivity, none of the community's leaders had been approached by government or non-government actors regarding the risks presented by climate change (Trundle and McEvoy 2015). Subsequent interviews with government stakeholders suggested that this was in large part due to the structural factors referred to in the first part

Figure 3.4 A Blacksands community chief provides an oral history of the impact of past climate events (left), including an explanation of historical erosion of the area's fertile volcanic soils (centre), visible in the dunes behind the Blacksands beach (right)

Source: Alexei Trundle Photo

of this chapter; namely, that much of peri-urban Port Vila fell between cultural and structural regimes (ibid.). Additionally, government and non-government actors alike expressed a view that climate change had not been previously understood to be 'an urban issue', and as such the infrastructure and social disadvantage in these areas does not feature strongly in Vanuatu's climate resilient development programmes and policies (GoV 2015a; GoV 2015b). Engagement in these informal, peri-urban spaces therefore built awareness within the community of the local impacts of climate change, while also legitimising climate adaptation actions and agency by local communities and other actors.

Workshops with provincial, municipal and national government representatives were used to convey the climate-related risks that these areas exhibit, as well as the nature of the changes that are likely to be expected into the future and that need to be planned for. Much of the vulnerability of peri-urban Port Vila, however, could be attributed to the existing 'adaptation deficit' evident in these domains (Bulkeley and Tuts 2013). For example, the lack of piped drinking water connectivity for these informal communities resulted in shallow wells being used for drinking water, which are prone to both saline ingress and contamination (in turn a product of the lack of sealed sanitation throughout the settlements). When examined through a climate vulnerability assessment methodology, this example demonstrates the multi-faceted nature of climate planning needs in similar informal zones. The use of climate hazard exposed land (in this case, coastal flood plains) is typical of informal settlements, which have limited ability to settle in lower-risk areas that are often already occupied. Their heightened climate sensitivity – measured, for example, by their lower income levels and poor physical infrastructure – exacerbates the impacts of this exposure. Finally, although community social and informal structures can be strong, the communities of Blacksands had no formal avenue for engagement with government departments and decision-makers, limiting their adaptive capacity.

In such settings, scientific experts can do more than simply analyse the climate-related vulnerability exhibited by these domains. Beyond this passive, 'top-down' contribution, climate-planning practitioners have the potential to provide a bridge between informal peri-urban agency and institutions and those of urban governance and decision-making. The involvement of government stakeholders in the Blacksands transect walks, for example, led to representatives of the National Disaster Management Office exchanging mobile phone numbers with community leaders, enabling direct involvement in disaster response and management. In this sense, scientific experts engaged in climate change planning can improve their own impact in the areas that they work by being aware of their own agency as external participants within complex and at times fragmented governance systems.

Local agency: attribution vs. causality

As demonstrated above, the cross-referencing of community experiences with scientific climate data in Blacksands was shown to be an effective tool for enhancing institutional understandings of – and planning for – local impacts and vulnerabilities. However, the intersection of expert-held climate change science (specifically, long-term local data, and downscaled climate change projections) and local knowledge of environment conditions, past hazard impacts and empirically observed trends also provides insights into the importance of understanding and communicating causality in adapting to climate change. Specifically, differentiating between changes being driven by local environmental degradation and those being driven by global anthropogenic climate change can play an important role in motivating local action and empowering agency in communities who have negligible responsibility for the state of the global carbon budget. Conversely, building an awareness of the causal responsibility for climate change in the hazard-exposed communities of the Pacific is leading to new modes of global engagement with the perpetrators of climate change internationally; carbon-intensive developed nations. Recent examples of this include the blockading of Australia's largest coal port by a flotilla of protest boats (including some traditionally constructed craft from the Pacific), and the call for a moratorium on new coal mines by Pacific Island leaders.

An example of the need to properly understand causality of observable environmental change can be found in the changes observed in the Tagabe River, which flows through the Blacksands informal settlement area. With almost a third of Blacksands households reliant on shallow wells for drinking water and more than 10 percent using the river for bathing and washing, observed reductions in the river flow and water table were noted by participants to be a key threat to health and livelihoods (Trundle and McEvoy 2015, p. 31). During consultations, several community members speculated that links could be drawn between reduced rainfall and globally driven anthropogenic climate change. However, this hypothesis was neither consistent with trends in recent rainfall records (Trundle and McEvoy 2015, p. 6), nor projections modelled under a range of future climate change scenarios, as shown in Table 3.1 (BoM and CSIRO 2014, p. 339). Instead, further investigation in partnership with the community was able to identify upstream diversion of the river by commercial water users, and increased groundwater extraction, as likely causes of the decrease in river flow. This, in turn, provided an evidence-based platform for community-based actions and advocacy for the reversal of these negative, local anthropogenic pressures.

Despite this lack of causal connectivity to global anthropogenic climate change, actions to improve the condition of local environmental processes can reduce relative climate change vulnerability, as they have the potential to enhance the function of and access to ecosystem services, thus building the capacity of local communities. As a result, these actions are regularly implemented as part of climate resilient development initiatives (Brooks et al. 2011; Brown et al. 2012). Similarly, the 'adaptation deficit' evident in the infrastructure in these informal communities inevitably – and appropriately – takes priority at a community level over actions that focus solely on future impacts (Bulkeley and Tuts 2013). With per capita emissions in Vanuatu estimated to be

Table 3.1 CMIP5 seasonal rainfall projections for Vanuatu (5–95 percent range bracketed)

Projected rainfall change	*Season*	*2030*	*2050*	*2070*	*2090*
RCP 2.6 (very low emissions)	**Wet**	+2% (−5 to +3%)	+2% (−6 to +9%)	0% (−9 to +14%)	+1% (−7 to +13%)
	Dry	0% (−11 to +12%)	+1% (−8 to +13%)	−1% (−17 to +9%)	−2% (−15 to +10%)
RCP 4.5 (low emissions)	**Wet**	0% (−8 to +15%)	+1% (−9 to +9%)	+2% (−8 to +18%)	+1% (−13 to +13%)
	Dry	0% (−12 to +15%)	−1% (−13 to +11%)	−2% (−14 to +12%)	−1% (−25 to +14%)
RCP 6 (medium emissions)	**Wet**	+3% (−5 to +15%)	+2% (−7 to +11%)	+3% (−5 to +16%)	+3% (−11 to +22%)
	Dry	+2% (−6 to +13%)	2% (−11 to +16%)	+2% (−11 to +18%)	+5% (−9 to +21%)
RCP 8.5 (very high emissions)	**Wet**	+1% (−6 to +12%)	+1% (−9 to +13%)	+3% (−14 to +17%)	+5% (−13 to +30%)
	Dry	−2% (−10 to +8%)	−1% (−19 to +16%)	−1% (−21 to +17%)	+3% (−26 to +34%)

Source: BoM and CSIRO 2014, in Trundle and McEvoy 2015

at around 0.4 tonnes of CO_{2e} in 2013, local capacity to mitigate climate change is limited, despite concerted efforts by government to reduce dependency on diesel for electricity generation (UNDESA 2016). Climate-resilient development therefore provides a viable and logical avenue for engaging with the issue of climate change, beyond advocacy and diplomatic efforts in the international arena.

The role of non-institutional adaptive and coping capacity to deal with climate shocks and stressors was also very much evident at a local, community scale in Port Vila's peri-urban zones. As shown in Figure 3.5, while institutional limits in areas such as finance, services, critical infrastructure and emergency response training were perceived to be lacking, community and non-government structures and knowledge were found to be key assets in coping with climate-related hazards. In peri-urban communities such as Blacksands, non-institutional capacity was found to be a critical coping mechanism during the recent impacts of Tropical Cyclone Pam. The ability to shelter with kinship networks and local civil society organisations (such as the Wan Smol Bag theatre company, which runs social services and facilities in the Blacksands area), has been attributed for the low number of lives lost during Cyclone Pam, as well as relatively speedy recovery and rebuilding (Hollema et al. 2015). Similarly, the lower levels of structural damage observed for traditionally built *nakamals* (a shelter that forms a *kastom* meeting area for the community) was noted as one example of the potential of local coping capacity that was underutilised in the makeshift and improvised buildings across Port Vila's informal areas (The World Bank 2015).

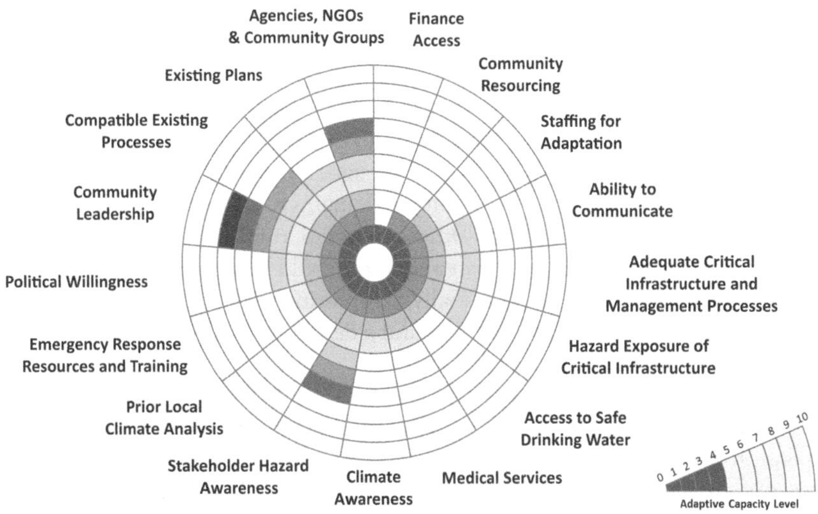

Figure 3.5 Stakeholder assessment of Port Vila's adaptive capacity (modified from Trundle and McEvoy 2015)

Urbanisation as adaptation?

Although 'resilience' is often promoted as an effective boundary concept for climate planning in cities (see, for example, Olsson et al. 2015), the interplay of urban functionality with external trade, resource and information networks is critical, and can act as a major source of shocks and stressors to the urban system (Tyler and Moench 2012). If the primary purpose and function of a city is understood as being to house and provide for its inhabitants, it is equally important to note that rapid population growth can in and of itself act as a shock on the system if poorly managed. At worst, rapid urban expansion has the potential to lead to disruption or system collapse, as was observed during the early 2000s in Honiara, during the ethnic violence that devastated the capital of Vanuatu's neighbour the Solomon Islands (MDPAC 2013). The extent to which urbanisation can be an effective mode of climate change adaptation at a national scale in Vanuatu is therefore poorly understood, particularly given the existing dysfunction evident in the governance of peri-urban Port Vila, which now constitutes more than 45 percent of the capital's growing population.

As noted by Bedford in his analysis of the 1979 census, urbanisation as a form of in-migration has long been a response to environmental shocks and stressors in Vanuatu:

> Another factor contributing to rapid urban population growth on both Efate and Santo in the early 1970s was a crisis in the rural economy . . . extensive cyclone damage to coconut groves in the densely populated Banks Islands

> and Shepherd Islands in 1971, coupled with a sharp fall in the price for copra in the same year, resulted in considerable rural-urban migration.
>
> (Bedford 1979, p. 75)

In a more recent study of the impact of the 2007 Global Financial Crisis, Fenny and McDonald also found that, perversely, urban inhabitants had a higher exposure to all forms of shocks due to a dependency on global markets (2013, p. 6), driven in part by a need to access non-urban domains to supplement food through bush gardens and plantation work. Other analysis of rural communities in Vanuatu also identified that rural-to-urban migration could be seen to have a negative overall effect on vulnerability nationally, by creating reinforcing feedback loops such as a dependence on imported food and a deterioration in traditional village structures in home islands (Craven 2015).

The ongoing trend of urbanisation in Vanuatu, while conceptually resisted and often under-emphasised in policy and governance, is nonetheless occurring as a response to a number of structural stimuli, within which the impacts of climate change are increasingly becoming intertwined. As critical thresholds for the collapse or transformation of ecosystems (particularly coral reefs and fisheries) are crossed over the next two to three decades (PACCSAP 2014), it is likely that urbanisation – as a function of the transition from subsistence to alternative sources of livelihoods – will occur at an increasing rate. However, as with most other impacts of climate change and all forms of migration, identifying climate change adaptation as a singular causal driver will be both uncommon where this is the case (through factors such as sea level rise on outer coral atoll islands) and unlikely given the additional factors at play. Nonetheless, climate change will continue to contribute to and accelerate already rapid rates of urbanisation in Port Vila.

Conclusion

Port Vila's peri-urban settlements have an unusual profile of climate change vulnerability, with significant levels of climate exposure and sensitivity being somewhat offset by high levels of community adaptive capacity and non-institutional, socially derived coping mechanisms. These attributes, as well as actionable local knowledge of these peri-urban areas' hazard profiles, are not evident from a 'top-down' assessment approach, but instead emerge through the application of participatory climate-planning processes such as that applied through the UN-Habitat *Planning for Climate Change* framework (UN-Habitat 2014). The integration of these community-based findings with climate science and data was found to be of mutual benefit for both community and institutional stakeholders, and a key vehicle for enabling local actors to improve their climate resilience.

Although the characterisation of Port Vila's peri-urban dwellers and inhabitants challenges many aspects of conventional understandings of the rural-to-urban transition process, as well as its interplay with climate change adaptation and planning, it is by no means unique to Vanuatu. Similar experiences of urban 'otherness' and migrant duality have been observed elsewhere – experiences noted to present unique perspectives on both individual cities and their theoretical framings, particularly postcolonial nationalism and the evolution of nascent forms of the nation state (see, for example, Lawson 2000). Challenging the exclusion of these informal spaces and settlers from urban institutions requires an in-depth understanding of their historical and cultural origins, as well as the local non-institutional arrangements that allow them to persist.

Tropical Cyclone Pam was estimated to have cost 64.1 percent of Vanuatu's Gross Domestic Product (The World Bank 2015), and with an average of 2.4 tropical cyclones passing through Vanuatu's exclusive economic zone per year (PACCSAP 2014), the need to address these immediate concerns through the integration of short-term disaster risk management and future-focused climate change adaptation is self-evident. Despite a lack of model agreement in either the direction or size of change to tropical cyclone frequency or intensity in the region under all future climate scenarios, the impacts of and response to Tropical Cyclone Pam demonstrate the rationale for shifting away from an emphasis on attribution and additionality toward climate-resilient development. In urban settings, this complements the complexities of inter-related local environmental pressures and human interventions with global anthropogenic climate change.

Finally, framing a city as a closed system, operating in isolation from its external social networks, global resource dependences and ecosystem-based interconnectivities can be conceptually limiting, as the case study of Port Vila's peri-urban fringe illustrates. While maintenance of arbitrary urban boundaries can provide a politically palatable option in the short term, it can severely limit the effective development of cities on longer time horizons, as highlighted by cross-boundary issues such as climate change.

Notes

1 *ni-Vanuatu* is the term used to describe the indigenous inhabitants of Vanuatu.
2 The figure of 44,039 is derived from the urban-classified Enumeration Areas that were used by the Vanuatu National Statistics Office in the 2009 Census (see VNSO 2010).

References

Banes, C., and Wilson, L. (2011). *Fact Finding Study of Urbanization in Vanuatu*. Final Report To AusAID, Australian Government, Canberra.

Bedford, R. (1979). *Population of Vanuatu: Analysis of the 1979 Census*. Noumea, New Caledonia: South Pacific Commission.

Brooks, N., Anderson, S., Ayers, J., Burton, I., and Tellam, I. (2011). *Tracking Adaptation and Measuring Development* (IIED Climate Change Working Paper Series No. 1). London. Retrieved from http://pubs.iied.org/pdfs/10031IIED.pdf?

Brown, A., Dayal, A., and Rumbaitis Del Rio, C. (2012). From practice to theory: Emerging lessons from Asia for building urban climate change resilience. *Environment and Urbanization*, 24(2), 531–556. Retrieved from http://doi.org/10.1177/0956247812456490

Bulkeley, H., and Tuts, R. (2013). Understanding urban vulnerability, adaptation and resilience in the context of climate change. *Local Environment*, 18(6), 646–662. Retrieved from http://doi.org/10.1080/13549839.2013.788479

Bureau of Meteorology (BoM) & Commonwealth Scientific and Industrial Research Association (CSIRO). (2014). Vanuatu country report update – Climate futures. In: *Climate Variability, Extremes and Change in the Western Tropical Pacific: New Science and Updated Country Reports*. 1st ed. Melbourne: CSIRO Publishing, 319–340. Retrieved from www.pacificclimatechangescience.org/wp-content/uploads/2014/07/PACCSAP_CountryReports2014_Ch16Vanuatu_WEB_140710.pdf

Chung, B.M., and Hill, D. (2002). *Urban Informal Settlements in Vanuatu: Challenge for Equitable Development*. Port Vila. Retrieved from www.forumsec.org/resources/uploads/attachments/documents/Vanuatu Squatter Settlement Study 2002.pdf

Craven, L.K. (2015). Migration-affected change and vulnerability in rural Vanuatu. *Asia Pacific Viewpoint*, 56(2), n/a–n/a. Retrieved from http://doi.org/10.1111/apv.12066

Ensor, J. (2015). *Adaptation and Resilience in Vanuatu*. Carlton, Australia. Retrieved from www.sei-international.org/mediamanager/documents/Publications/Climate/Oxfam-SEI-2016-Vanuatu-adaptation-resilience.pdf

Feeny, S., and McDonald, L. (2013). Multidimensional poverty and vulnerability in Solomon Islands and Vanuatu. In: Fenney, S. (ed.) *Addressing Inequality and Promoting Inclusive and Sustainable Development*. Suva: Ashgate Publishing, 1–23. Retrieved from http://mams.rmit.edu.au/d0c5gfhi3otl.pdf

Government of Vanuatu (GoV) (1980). *Port Vila Municipality (Delineation of Boundaries) Declaration* [CAP. 126]. Government of Vanuatu.

Government of Vanuatu (GoV) (1986). *Report of the Vanuatu Urban Census 1986*. Port Vila: National Planning and Statistics Office.

Government of Vanuatu (GoV) (1987). *Second National Development Plan 1987–1991*. Port Vila: Vanuatu.

Government of Vanuatu (GoV) (1989). *Vanuatu National Population Census – May 1989*. Port Vila: National Planning and Statistics Office.

Government of Vanuatu (GoV) (2013). *Vanuatu National Land Use Planning Policy*. Port Vila: Vanuatu.

Government of Vanuatu (GoV). (2015a). *Vanuatu Climate Change and Disaster Risk Reduction Policy 2016–2030*. Port Vila, Vanuatu: Secretariat of the Pacific Community. Retrieved from www.nab.vu/sites/all/files/vanuatu_cc_drr_policy_minus_att4v4.pdf

Government of Vanuatu (GoV) (2015b). *Vanuatu Infrastructure Strategic Investment Plan*. Port Vila: Vanuatu. Retrieved from www.theprif.org/components/com_jomcomdev/files/2015/08/33/119-VISIP 2015–2024 Report web.pdf

Hardwick, M. (2012). *Blacksands, Efate – Community Profile. Vanuatu Community Resilience Project*. Shefa Provincial Government. Port Vila, Vanuatu.

Heron, D., Lukovic, B., Smart, G., Ramsay, G., Roberts, G., Hopkins, L., and Todman, S. (2016). *New Approaches to Hazard and Risk Mapping, Vanuatu – Developing Safer Communities*. Retrieved from http://star.gsd.spc.int/meeting_docs/presentations/Session7-2a_Hazard&RiskMappingVanuatu_HeronD.pdf

Hollema, S., Miller, D., and Chong, A. (2015). *Vanuatu: Cyclone Pam Impact Maps & Analysis*. Port Vila, Vanuatu. Retrieved from www.humanitarianresponse.info/fr/system/files/documents/files/Vanuatu WFP-VAM assessment-final.pdf

Johnson, J., Bell, J., and Gupta, A. (2015). *Pacific Islands Ocean Acidification Vulnerability Assessment*. Secretariat of the Pacific Regional Environment Programme. Apia, Samoa.

Lawson, V.A. (2000). Arguments within geographies of movement: The theoretical potential of migrants' stories. *Progress in Human Geography*, 24(2), 173–189. Retrieved from http://doi.org/10.1191/030913200672491184

Leslie, E. (2012). *Making a Living in Vanuatu: Livelihoods and Development in Peri- Urban Port Vila.* University of Otago. Retrieved from https://ourarchive.otago.ac.nz/bitstream/handle/10523/4461/LeslieEmilyM2013MPlan.pdf?sequence=3

Lightner, S., and Naupa, A. (2005). *Histri Blong Yumi Long Vanuatu: An Educational Resource.* (Delbé, M. and Lacrampe, S. Eds.) (1st ed., Vol. 2). Port Vila: Vanuatu Cultural Centre.

Mcdonnell, S. (2013). Exploring the cultural power of land law in Vanuatu: Law as a performance that creates meaning and identities. *Intersections: Gender and Sexuality in Asia and the Pacific*, December, 33, 1–17.

McEvoy, D., de Ville, N., Komugabe-Dixon, A., and Trundle, A. (2016). *Greater Port Vila Ecosystem and Socio-Economic Resilience Analysis and Mapping (ESRAM) – Technical Summary*. Report for the Secretariat of the Pacific Regional Environment Programme (SPREP). Apia, Samoa.

Mecartney, S.N. (2000). *Blacksands Settlement: A Case for the Urban Permanence in Vanuatu.* Sydney: University of Sydney.

Ministry of Development Planning and Aid Coordination (MDPAC) (2013). *Solomon Islands National Infrastructure Investment Plan.* Solomon Islands Government, Honiara.

Mucke, P. (2015). World Risk Report 2015. *Stuttgart.* Retrieved from http://collections.unu.edu/eserv/UNU:3303/WRR_2015_engl_online.pdf

National Housing Corporation (NHC). (2012). *Port Vila Informal Settlements Upgrading Project Basic Data on Port Vila Informal Settlements.* Government of Vanuatu. Port Vila, Vanuatu.

Olsson, L., Jerneck, A., Thoren, H., Persson, J., and O'Byrne, D. (2015). Why resilience is unappealing to social science: Theoretical and empirical investigations of the scientific use of resilience. *Science Advances*, 1(4), e1400217–e1400217. Retrieved from http://doi.org/10.1126/sciadv.1400217

Pacific-Australia Climate Change Science and Adaptation Planning Program (PACCSAP) (2014). *Pacific Climate Change Data Portal – Vanuatu Historical Climate Information.* Online Resource, Retrieved from www.bom.gov.au/pacific/vanuatu/

Rawlings, G.E. (1999). Foundations of urbanisation: Port vila town and Pango village. *Vanuatu Oceania*, 70(1), 72–86. Retrieved from http://doi.org/10.1002/j.1834-4461.1999.tb02990.x

Shorten, G., Schmall, S., Caron, A., Goosby, S., Granger, K., Naidu, P., and Titov, V. (2004). *Disaster Risk Management in Marginal Communities of Port Vila, Vanuatu.* Port Vila. Retrieved from www.pacificdisaster.net/pdnadmin/data/original/sopac disaster risk management in marginal communities of port vila.pdf

Tawney, E.J. (2006). *Watershed Management and Planning in the Tagabe River Watershed Catchment Area in and Around Port Vila, Vanuatu.* Michigan Technological University, Michigan, United States. Retrieved from www.mtu.edu/peacecorps/programs/civil/pdfs/eric-tawney-thesis-final.pdf

Trundle, A., and McEvoy, D. (2015). *Greater Port Vila Climate Change Vulnerability Assessment.* United Nations Human Settlements Programme (UN-Habitat), Regional Office of the Asia Pacific. Fukuoka, Japan.

Tyler, S., and Moench, M. (2012). A framework for urban climate resilience. *Climate and Development*, March, 4, 311–326. Retrieved from http://doi.org/10.1080/17565529.2012.745389

United Nations Department of Economic and Social Affairs (UNDESA). (2013). *World Statistics Pocketbook 2016 Edition.* New York. Retrieved from http://unstats.un.org/unsd/publications/pocketbook/files/world-stats-pocketbook-2016.pdf

United Nations Human Settlements Programme (UN-Habitat) (2012). *Streets as Tools for Urban Transformations in Slums: A Street-Led Approach to Citywide Slum Upgrading*. United Nations Human Settlements Programme. Nairobi, Kenya.

United Nations Human Settlements Programme (UN-Habitat) (2014). *Planning for Climate Change: A Strategic, Values-Based Approach for Urban Planners – Toolkit*. United Nations Human Settlements Programme. Nairobi, Kenya.

Vanuatu National Statistics Office (VNSO) (2010). *Vanuatu 2009 National Population and Housing Census – Basic Tables Report*, Volume 1. Port Vila, Vanuatu. Retrieved from www.pacificdisaster.net/pdnadmin/data/original/VUT_2009_Census_ReportV1.pdf

Verisk Maplecroft (2015). *5th Annual Natural Hazard Risk Atlas*. Verisk Analytics. Bath, UK.

Woodward, K. (2014). A *Political Memoir of the Anglo-French Condominium of the New Hebrides Arrival: Port Vila, Santo and the Condominium (eBook)*. Canberra: ANU Press. Retrieved from http://press.anu.edu.au/wp-content/uploads/2014/10/ch013.pdf

The World Bank. (2009). *Reducing the Risk of Disasters and Climate Variability in the Pacific Islands – Republic of Vanuatu Country Assessment*. Retrieved from www.sprep.org/att/irc/ecopies/countries/vanuatu/113.pdf

The World Bank. (2015). *Post-Disaster Needs Assessment: Tropical Cyclone Pam*, March, 172. Retrieved from http://dfat.gov.au/about-us/publications/Documents/post-disaster-needs-assessment-cyclone-pam.pdf

4 'Learning by Doing'

Lessons from the co-production of three South African municipal climate change adaptation plans

Julia Davies and Gina Ziervogel

Introduction

The relative novelty of the climate adaptation field and the context-specific nature of adaptation means that there are no readily transferrable frameworks to guide adaptation planning. As a result, governments have largely adopted an experimental, 'learning-by-doing' approach (Anguelovski et al. 2014; Bulkeley and Broto 2013; Roberts et al. 2012). Some of these adaptation processes have been sparked by the need to build adaptive capacity and increase resilience to natural disasters, whilst others have been linked to external pressure (e.g., from international climate policy commitments) and the desire by officials to achieve certain departmental agendas or to demonstrate leadership ability (Anguelovski et al. 2014). Adaptation planning in other cases has been in response to opportunities to receive support and technical guidance from international donor organisations or the private sector, for example to implement projects such as green roofs for urban stormwater management (Chu 2015; Mees et al. 2015). However, many are initiated internally and sustained by endogenous forces (Carmin et al. 2012). For example, in Durban, adaptation was department-led and driven by an individual champion who was able to access international funding (Roberts 2008), and in Quito, adaptation was initiated by the efforts of the municipality whose leaders recognised the vulnerability of the city's rapidly growing population to increasing glacial melt (Anguelovski et al. 2014).

While a greater level of insight into these iterative processes is being developed, this has generally been in a localised sense and only a limited number of studies have assessed these diverse approaches through a comparative lens. This means that there is still a lack of understanding around local adaptation trends and emerging forms of adaptation governance. One pattern that has developed is the tendency for local governments to draw on more inclusive and participatory forms of climate change response planning (Chu et al. 2015). The practice of adaptation planning has increasingly been matched with discourses around collaborative governance, in which stakeholder participation and knowledge co-production are key (Ansell and Gash 2008; Clarke et al. 2013; Leck and Simon 2013). Whilst both the merits and challenges of collaborative planning processes have been well documented (Barton et al. 2015; Chu et al. 2015; Cloutier et al.

2014), less is known about how diverse actors are drawn into these processes, what role they play and whether collaboration itself has transformative potential (for individuals or organisations) that endures beyond the planning phase.

This chapter assesses, comparatively, the adaptation-planning processes that were undertaken in the Bergrivier, Drakenstein and Eden Municipalities. These municipalities are situated in the Western Cape Province of South Africa (SA), which is a high-risk region in terms of both climate impacts and socio-economic vulnerability (WCG 2014; Ziervogel et al. 2014). Poor and marginal communities in the province tend to be the most affected by climate impacts. Yet, the municipal governments that are tasked with serving them[1] are often unable to do so, as their focus is on municipal-wide infrastructure and services rather than strengthening local capacity. A large portion of the municipal budget comes from local rates, which can be low if there is a high proportion of poor households. Human capacity is also a challenge, as many officials are overworked and do not have the skills to develop and implement climate change responses. Such challenges tend to be complicated by the environment in which local governments operate, wherein there are multiple issues that need to be prioritised in addition to climate change adaptation, the latter of which is not mandated by formal climate change legislation (Clifford-Holmes et al. 2015; Roberts and O'Donoghue 2012; Spires 2015; Ziervogel and Taylor 2008).

In recognising the need to aid municipalities in adapting to climate change, the Western Cape Government (WCG) introduced a Climate Change Municipal Support Programme (CCMSP) in 2012. This is an ongoing initiative that assists municipalities in developing climate change response plans through a series of multi-stakeholder engagements. These engagements take different forms, including workshops and focus groups, but have primarily been dominated by government officials, with some areas bringing in a range of actors from the private sector and civil society. The overall vision of this collaborative process is to mainstream climate change adaptation into development planning at the local level to allow adaptation imperatives to be aligned with the priorities of existing government policies. This is achieved when climate change adaptation is embedded in strategic documents such as Integrated Development Plans (IDPs) and Spatial Development Frameworks (SDFs). Mainstreaming is important not only because of the financial and capacity constraints that prevent a dedicated climate change response at the municipal level, but also because it is these master planning documents which hold legislative force (Pasquini et al. 2013).

This chapter aims to identify the key lessons learnt from the climate adaptation planning processes of three municipalities. It begins by describing the methods used to conduct the study, before providing the context and details of the adaptation planning processes. Three lessons that were learnt from these processes are then presented.

Methods

The Bergrivier (BRM), Drakenstein (DM) and Eden (EDM) Municipalities were selected as case studies for this chapter as they were the first to participate in the adaptation component of the provincial government's CCMSP, which took place

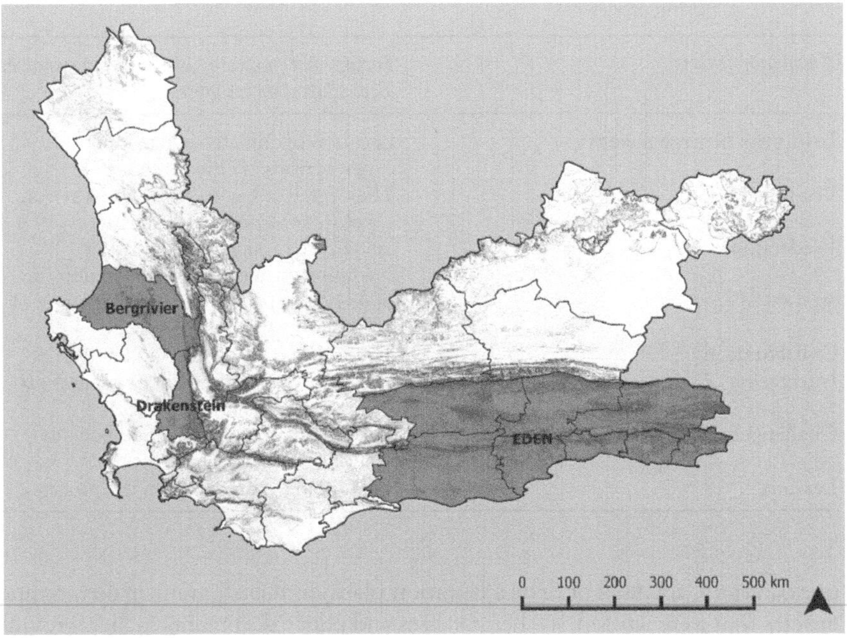

Figure 4.1 Map of the Western Cape Province showing the location of the Bergrivier, Drakenstein and Eden Municipalities.

Source: Author's own, developed using QGIS (2016)

between 2012 and 2013. According to the Municipal Structures Act of 1998 (RSA 1998a), BRM and DM are classified as Category B local municipalities, whereas EDM is a Category C district municipality. Category B municipalities are the smallest sphere of local government and are located outside of major metropolitan areas (Category A municipalities). A number of B-municipalities (usually around 4–6) make up one district (Category C) municipality, and development and service delivery in the area is a shared function of both. Figure 4.1 indicates the location of the three municipalities within the Western Cape Province.

Data for the study was collected primarily through fourteen semi-structured interviews, which were conducted with stakeholders over November and December 2015. Four of these were with representatives from the BRM adaptation planning process, three were from the DM process, four from the EDM and three from the WCG.

The interviewer was not present during the adaptation planning processes, and interviewees were therefore asked to reflect on their experiences. Interviews were conducted in a flexible manner but questions were generally guided by the topics highlighted in Table 4.1.

Secondary data was collected through ad-hoc analysis of documents such as the WCG's invitation to municipalities to take part in the CCMSP, their responses and letters of acceptance from the WCG, meeting minutes, workshop

Table 4.1 Guiding topics for semi-structured interview questions on municipal adaptation planning processes

Adaptation drivers	*Factors that motivated, enabled and sustained local adaptation responses*
Institutional arrangements	Location of climate change function, perceptions of climate change
Legitimacy	The degree of buy-in to the process from politicians and stakeholders
Leadership	Level/type of support provided by administrative leaders and champions
Actors	Stakeholder participation and the role of different actors in the process
Collaboration	The role of relationships and networks
Integration	Mainstreaming/alignment of adaptation with existing activities
Challenges	The overall challenges of and benefits derived from each process
Lessons	'Take-home' messages from the process

presentations, the final or draft adaptation plans from each municipality, online articles that were written on the processes and policy documents such as provincial and municipal IDPs.

The qualitative data from the interviews was assessed comparatively. Following Taylor et al. (2014) the comparative approach used was relational across the three municipalities, as opposed to evaluating each case in terms of its degree of conformance to a universal ideal. The analysis was less concerned with the specific content of the adaptation plans and the implementation outputs thereof. Rather, the focus was around the manner in which they were produced, the role and influence of different actors and their voices in informing these plans, and the various factors both driving and sustaining the municipalities' adaptation responses.

Adaptation planning in the Bergrivier, Eden and Drakenstein municipalities, South Africa

Bergrivier municipality (BRM)

Bergrivier is local municipality situated within the West Coast District Municipality. It covers an area of 4407 km^2 and in 2011 had an estimated population of 61,897 (Stats SA 2012). The municipality's adaptation planning process was carried out in four main stages, including a preliminary meeting between key representatives from the BRM and the CCMSP team, followed by three workshops. Participants at the workshops included the mayor, the municipal manager, the manager of strategic services (who took on the role of climate change champion), local and district municipal officials, the local director of tourism and disaster manager, officials from the conservation sector and private sector representatives, as well as local residents, including farmers and a group of Rastafarians. The number of attendees at the three workshops ranged from 20 to 35.

The BRM's adaptation planning process included two additional partnerships beyond the province. Two academics, including one of the authors of this chapter, with significant experience in climate adaptation and vulnerability, approached the WCG Climate Change Directorate about the possibility of working with a municipality to look at how climate information might be used at the local level. The Climate Systems Analysis Group (CSAG) at UCT, with their experience in climate services, also joined the team.

The combination of various skills in the BRM team allowed for flexibility in response to context-specific situations and emergent information. For example, when one of the workshop activities was found not to be working in the way that the facilitators (that included academics and the lead provincial stakeholder) had hoped, they were able to use their knowledge and experience gained from other workshopping processes to adjust the activity to one more suited to the context. The benefits of having a multidisciplinary team were further enhanced by the presence of a willing and receptive municipal environment. Fundamental to this was strong political support and an enthusiastic municipal climate change champion who had a passion for the environment. Although the municipality had not engaged in climate change issues directly before, the idea was presented to the Bergrivier Council, who gave their political support. The champion demonstrated outstanding enthusiasm and came to be viewed by the CCMSP team as essential to the success of the process. She was described as "*this amazing thing . . . a very unique individual*" (Respondent D, researcher). The champion was able to recognise her own limitations in terms of her climate knowledge, and instead apply her skills as a facilitative leader to organise and drive the process. She was successful in drawing an array of stakeholders into the planning process and thus demonstrated integrative leadership skills (Crosby and Bryson 2010). In part, this leadership was due to the small size of the municipality and its coherent nature, where if a senior person took initiative, there was support for her work.

Bergrivier, like many local municipalities in SA, have little financial capital to allocate to areas outside of mandated municipal functions, such as service delivery and local economic development, and climate change adaptation is commonly seen as an additional functional without budget. Mainstreaming was therefore identified in the BRM Climate Adaptation Plan (BRM 2014) as one of the key areas for adaptation intervention in the municipality, the approach being to consider climate compatibility as fundamental to the municipality's development planning rather than as an added extra.

Whilst the municipality has faced some difficulties in implementing the plan, it was felt by most of the stakeholders that engaging in the adaptation planning process itself was highly valuable. The experience resulted in a significant degree of learning among the officials, as well as capacity-building for local stakeholders. Unfortunately, given that there is no climate change mandate, additional resources did not accompany the plan, and so there was no focused implementation of projects. In addition, the champion left the municipality and the new strategic manager focused on other issues such as local economic development, rather than climate adaptation.

The provincial representatives and experts were also able to bridge the interface between science and policy in a constructive way: "*it's a 'win-win' situation – we are able to learn so much about things from a government/policy perspective, and can then feed this into our work*" (Respondent C, researcher). The process led to the foundation of new networks among the participants, as well as to the strengthening of existing relationships between those who had worked together in the past. The partnership with UCT also provided the opportunity to creatively overcome resource, capacity and knowledge deficits by providing research projects for postgraduate students around climate change vulnerability and adaptation in the municipality.

Drakenstein municipality (DM)

Drakenstein Municipality covers a municipal area of 1,538 km^2 and is home to an estimated 251,262 people, according to Census 2011 (Stats SA 2012). DM forms one of five category B municipalities in the Cape Winelands District. DM's adaptation planning process began in October 2012 with an initial inception meeting between the provincial CCMSP team and various key representatives from the municipality, followed by a stakeholder workshop. The workshop was attended by a group of 16 participants who were primarily municipal officials from various different line functions, including those such as Community Services, Financial Services, Civil Engineering Services, the Department of Environmental Management and the Department of Water Services. Emerging from the consultations with these stakeholder groups was a variety of sector-specific interests, which had considerably divergent objectives. Thus, whilst the programme was originally planned to include participatory workshops, the process was adapted to include targeted focus group meetings instead.

The climate change champion for the DM, who led the process, was the municipality's senior engineer, who at the time was concurrently responsible for issues related to environmental management. His keen interest in environmental affairs led him to play a key role in the inclusion of environmental management in the municipality's existing functions.

Despite the municipality's cognizance of environmental issues and climate change, there were a number of barriers that inhibited an optimal adaptation planning process from occurring, a consequence of which was that the plan was not fully completed. One of the biggest difficulties was that at the time of undertaking the CCMSP process, the municipality was undergoing a significant restructuring, having recently come under new administrative management. This caused various problems to arise, including communication breakdowns and tension between departments: "*this is now what the big fight is about – how are we going to separate the planning portion of environmental management from the operational? Personally, I can't see how that can be done*" (Respondent G, DM).

The DM's process also lacked sufficient buy-in from political and administrative heads. Although there was no top-down resistance to developing an adaptation plan for the municipality, the general perception of climate change was as a 'green' issue, and it thus lacked emphasis in relation to more immediately pressing development concerns, such as unemployment, crime and housing shortages. This problem was expressed by a government official, who observed that the climate change champion lacked support:

> [He] tries, but I don't think he gets the same . . . either political or senior administration ear, you know – the people holding the purse strings, they 'rule the roost'. So he tries his best but I don't think it actually translates into projects, which is what he wants. So I don't think that climate change is seen as a priority.
>
> (Respondent A, WCG)

A further barrier to the process was that the stakeholder group participating was primarily composed of government officials, many of whom occupied junior positions within their respective departments. Whilst it is important in such a process to have representation from various sectors and departments, there was a significant absence felt from stakeholder groups such as senior management, industry, civil society and non-governmental organisations (NGOs). The diversity of stakeholders involved in the participatory engagements did improve slightly towards the later stages of the process. However, it seems that there was an initial misunderstanding in terms of who was expected to attend and what the imperatives of the adaptation plan were, despite the discussions that took place with the municipality prior to commencing the process. One of the provincial actors felt that this was partially due to the municipality's champion having a 'traditional' perception of environmental management as being separate from the broader development context, which includes climate adaptation as relating to housing, resource access, employment and other challenges.

The challenges experienced in the DM's adaptation planning process cannot be attributed to a lack of awareness about climate change or an unwillingness to implement projects. Rather, many of these challenges are related to what the actual drivers of or reasons for responding to climate change are, and the way in which the specific actions are framed. This is commonly from a perspective of disaster management or economic efficiency as opposed to climate change adaptation or mitigation:

> A lot of the things that are potentially climate-change based or linked, we've been doing for other reasons . . . the main drivers sort of come out of the social impacts from a disaster management perspective.
>
> (Respondent F, DM)

Despite the difficulties experienced during the adaptation planning process, both the municipal stakeholders and the provincial team were able to realise the interconnected nature of climate change adaptation and mitigation and the need to practice these response types simultaneously: "*as we were working through the whole thing we realised that there is so much overlap between adaptation and mitigation at a municipal level. If you do a mitigation project there, generally there are adaptation co-benefits attached to that*" (Respondent H, WCG). Adopting a merged approach would also help in overcoming the capacity constraints of the provincial team, whose members were split between separate adaptation and sustainable energy planning processes.

Eden district municipality (EDM)

In contrast to BRM and DM, which are local municipalities, Eden is a district municipality that includes seven local municipalities. EDM is situated on the southern coast of the Cape, covers an area of 23,331km^2 and in 2013 had a population of 587,564 people (WCG 2014a)

EDM's climate adaptation planning process was initiated in August 2012 when an introductory meeting was held in the municipal seat of George. The meeting was well attended by a range of key stakeholders, including the mayor, local councillors and various municipal officials. In light of the district scale at which this process was to be carried out, it was vital to secure sufficient buy-in to the programme not only from the administrative and political district heads but also from the local municipalities within the district. It was agreed that three subsequent adaptation planning workshops would be held.

Whilst the district-wide scale of the EDM process made it significantly challenging, the enthusiasm and strong leadership on the ground made for a highly receptive working environment in which significant learning and growth occurred. The EDM's environmental manager, who led the work, drew on his existing networks and relationships with the B-municipalities to engage stakeholders from various government departments as well as from NGOs and institutions such as SANParks and Cape Nature.[2] This broad stakeholder base allowed for heightened interactions and good dialogue to occur amongst the participants. Whilst the number of representatives at the workshops waned toward the end of the process and there was a lack of representation from sectors such as engineering and industry, the overall process saw the emergence of robust engagements which had been absent in the past.

Although the EDM Climate Adaptation Plan is still in draft status, a notable outcome of this process was that three of the local municipalities (including Knysna, Mossel Bay and Hessaqua) incorporated climate change adaptation into their local level IDPs. A co-ordinated response is important, as although the district has long been responding to climate impacts in various ways, there has been insufficient integration among various role players, with many plans and interventions occurring independently.

One of the biggest links that was missing in this regard was between the Environmental and Disaster Management units, which for a long time had been at loggerheads with regards to the prioritisation of issues and the approaches being taken to address climate-related impacts. However, after the occurrence of disasters such as the severe floods in 2007 and the 2009–2011 drought, the need for improved collaboration between the departments was realised. The adaptation planning process served as a significant catalyst in strengthening the relationship between the two spheres. This occurred on a planning level, as "*for the first time you've got a matrix that displays what everyone is busy with*" as well as on a cognitive level, with officials realising that "*we are all trying to reach the same goal . . . to maximise the net result*" (Respondent K, EDM).

Having worked with both EDM's Environmental and Disaster Managers on various projects and forums in the past, the lead provincial stakeholder became a key architect in enriching the interactions between the two departments. The stronger relationship that was formed resulted in the district's Disaster Manager emerging as an unofficial co-champion of the climate adaptation agenda. A sturdy and dynamic link between these sectors is essential not only in light of the disaster-prone nature of the Eden District (Nel et al. 2014), but also in terms of the legislative requirements around climate change and disaster risk reduction in South Africa, with only the latter of these being a compulsory obligation for municipalities.

Building and drawing on relationships and networks is a key aspect of effecting a valuable climate change response, particularly in terms of garnering the support that is needed to implement projects that take climate change into account. This was demonstrated at the district scale by improved collaborations between the environmental and disaster management units. This was further expressed by some of the interview participants from local municipalities for whom relationships mean greater access to knowledge and the ability to combine efforts and affect a more robust outcome. In this regard, informal knowledge channels such as text message groups are seen as being highly useful for extracting and sharing information, which in turn can lead to more effective climate change response planning and action. Similarly, it was found that building closer relationships with individuals in different departments is important for achieving outcomes at the local level:

> It helps to be friends with the guy . . . when you add something more to the professional relationship then you know you are getting something richer and more solid . . . a personal relationship helps to enhance the final product that you are trying to achieve.
>
> (Respondent I, EDM)

The adaptation planning process in EDM served as a platform through which climate change could be addressed horizontally, across different sectors and departments and vertically, as the provincial Climate Change Directorate engaged with stakeholders at the district scale, which subsequently allowed for adaptation responses to be filtered down to the B-municipalities who had also been involved in the district's adaptation planning process.

Key lessons from municipal climate adaptation planning in the Western Cape

The variable experiences of BRM, DM and EDM illustrate that there is no 'one-size-fits-all' approach to adaptation at the local level. Rather, adaptation planning has been characterised by experimentation and 'learning-by-doing' (Anguelovski and Carmin 2011; Carmin et al. 2009). Table 4.2 provides a summary of the major differences that arose during these three processes. By comparing the results of the case studies, we can identify important lessons that may help to improve existing approaches to adaptation policy and practice.

Lesson 1: political will and suitable local leadership are preconditions for success

A strong sense of political will and buy-in to the adaptation agenda from politicians has shown to be key for enabling more effective policy processes (Pasquini and Shearing 2014; Pasquini et al. 2014). Whilst it is possible to drive climate adaptation from within the administrative section of a municipality (such as in BRM), it is essentially in the hands of politicians that decisions around resource allocations lie. Thus, "engaging senior leadership and getting [political] 'buy-in from the top' is key for unlocking resources for, and coordinating approaches to, organisational adaptation" (Turner et al. 2016, p. 19). Without adequate buy-in to the process 'from the top', any efforts from municipal officials or departments to affect an adaptation response may be in vain. This was shown in DM, where inadequate buy-in from elected officials, combined with the failure of senior management to prioritise climate adaptation, meant that the agenda became somewhat lost when the municipality underwent a restructuring process.

Strong support from senior officials within the municipality is also essential. This was demonstrated in BRM, where political buy-in was complemented by the value attributed to adaptation by the Municipal Manager and the Manager of Strategic Services. Similarly, in EDM, the emphasis placed on responding to climate change by district leaders filtered down to some local municipalities, who subsequently included adaptation in their local-level IDPs. Significant variances across the three cases in this study indicate that the degree to which such positive outcomes are achieved is strongly dependent on a set of nuanced factors, including the nature of the individual climate change champion in question. It is therefore not only the presence of champions that are important but the *type* of leadership qualities that they possess.

The value of certain individual qualities was exemplified in BRM, whose champion illustrated how a local leader, even in the absence of in-depth knowledge about climate change, can enable an effective adaptation response. Motivated by her enthusiasm for the topic, she was not constrained by factors such as a lack of resources, but used her existing skill set, and humble but persuasive nature, to draw participants into the workshops and facilitate a collaborative process. One important quality was her willingness to co-operate and fully engage with

Table 4.2 Key differences between the BRM, DM and EDM municipal adaptation planning processes

	Bergrivier municipality	*Drakenstein municipality*	*Eden municipality*
Municipal scale	Local	Local	District
Status of plan at time of study	Finalised and approved in 2013, adopted in 2014.	Incomplete draft	Final draft completed in 2014
Institutional location of climate change	Office of the Municipal Manager: Directorate of Strategic Services (included under 'environmental management' function)	Included in environmental management function under Directorate of Planning & Economic Development and Department of Civil Engineering (separation of planning and operational components of environmental management)	Department of Environmental Management and Department of Disaster Management
Internal legitimacy	Good buy-in from political and administrative leaders – enabled partly by champion who was the strategic manager	Although no opposition to adaptation from leaders, there was a lack of support/buy-in from politicians	Support from political and administrative leaders. Lack of buy-in/commitment in some B-municipalities
Leadership	Champion was a highly effective facilitative and integrative leader	Champion initiated an environmental focus in the Municipality but held traditional views around environment vs. development	Enthusiastic and committed champion. Strong leadership emerged from disaster management: co-championing across sector departments enabled improved alignment of goals, shared resources etc.
Structure of planning process	Inception meeting and three multi-stakeholder workshops	Inception meeting, one multi-stakeholder workshop and one workshop session of targeted focus group interviews	Inception meeting, three multi-stakeholder workshops (inclusive of all B-municipalities) and multi-stakeholder workshops with four individual B-municipalities
Stakeholder participation at workshops	20–35 participants. Broad stakeholder representation, including the Mayor, senior officials, local residents, conservation sector etc. Greater representation from farmers would have been beneficial	12–16 participants. Lack of stakeholder diversity and continuity. Mainly junior municipal officials from various different line functions. Lack of input from senior management, industry, civil society and NGOs. Stakeholder diversity improved slightly toward later stages of adaptation planning process	9–16 Participants. Broad stakeholder representation including Mayor, senior officials, NGOs etc. Number of participants waned toward end of adaptation planning process and there was a lack of representation from sectors such as engineering, industry and civil society

(*Continued*)

Table 4.2 (Continued)

	Bergrivier municipality	*Drakenstein municipality*	*Eden municipality*

Relationships and networks	Existing relationships among members of the core team presented opportunities to build partnerships. Involvement of key partners allowed for multidisciplinarity and co-production of knowledge	Municipal restructuring caused shift in stakeholder responsibilities and communication breakdowns between departments. Strong ownership of climate change by Municipal champion and lack of existing relationship between stakeholders and WCG created gap in communication channel. Divergent ideas/perceptions among different stakeholder groups led to more focused group interviews – resulted in lack of knowledge co-production and poor dynamism	Champion and provincial representative had worked together in the past. Champion had good relationship with B-municipalities and strong network ties with organisations in environmental/ conservation sector. Strengthened relationship between environmental and disaster management departments
Flexibility/ innovation	Planned workshop activities adapted to improve relevance and effectiveness based on emergent needs and information. Opportunity to creatively overcome resource, capacity and knowledge deficits by providing research projects for postgraduate students – enabled by partnership with UCT	Workshop structure adapted to accommodate the need for more focused group meetings. Framing climate change as a disaster risk response or local economic development imperative enables greater political acceptance and support	Additional round of workshops held with individual B-municipalities to allow for more refined adaptation planning support. Informal communication channels enable networking and personal relationships enable easier exchange of information. Unofficial co-championing of climate change by Disaster Manager allowed for more integrated and proactive adaptation response.
Social learning	Champion open to new experiences and willing to learn from others. Workshops helped to bridge science-policy interface. Individual learning led to capacity building for community members	Realisation of strong ties between adaptation and mitigation at local level resulted in shift in approach towards more holistic climate change response	Multi-stakeholder workshops provided opportunity to realise ties between the work of different departments and enabled officials to realise that they are pursuing a common goal

Mainstreaming	Mainstreaming prioritised as key adaptation intervention. Adaptation plan rapidly finalised and timed to coincide with annual IDP review	Incompletion of plan resulted in missed opportunity to mainstream adaptation agenda into 2014/2015 update of the Municipality's SDF.	Three of the local municipalities incorporated some of the measures from the District's draft adaptation plan into their local level IDPs

the multidisciplinary team, rather than take control of the process. Conversely, whilst the DM champion was enthusiastic about developing an adaptation plan, his 'entrained thinking' from his previous work meant he was less flexible in adapting to new approaches (Snowden and Boone 2007).

This suggests that greater emphasis should be placed on identifying champions who not only have an interest or stake in the climate change agenda, but who have demonstrated facilitative and integrative leadership skills, who are humble, who are willing to learn continuously, who embrace diverse perspectives and who accept change. The implications of this policy lesson suggest the need to address not only the 'visible' barriers to adaptation, but also the underlying factors that cause adaptive barriers to arise.

Lesson 2: a collaborative and integrated approach is required

Collaboration and knowledge co-production are key to the design of the CCMSP. However, the degree to which this objective was achieved in the three municipalities varied according to factors such as the emergence of opportunities for collaboration, the presence of existing relationships and networks, as well as the receptiveness of municipal stakeholders to collaborative engagements. In terms of the latter, a failure in DM to involve a broader group of non-state actors and senior officials in the process was linked to a strong custody of the climate change mandate by officials in the environmental sector. Whilst ownership of the issue is important, in this context it acted as a barrier to realising greater integration between actors.

The adaptation planning process undertaken in BRM can be viewed almost as the antithesis to that of Drakenstein. The Bergrivier Municipality was afforded a number of capacity-building opportunities that were made possible through the incorporation of climate scientists and academic researchers into the process. These partnerships allowed for a marriage between scientifically rigorous and policy-relevant information, and enabled the establishment of strong network ties amongst the various actors that built on previous networks and relationships.

The incorporation of state and non-state actors from various sectors and disciplines is an important aspect of knowledge co-production and partnership-building (Swilling 2014), as illustrated in the BRM. Additionally, sustainable responses to complex issues such as climate change require that both vertical and horizontal governance networks are formed (Amundsen et al. 2010; Henstra 2015). Whilst there was an initial lack of integrated, cross-sectoral planning in EDM, the collaborative nature of the adaptation planning process allowed a closer relationship to be formed between the Environmental and Disaster Management Departments. By maintaining links between different departments across multiple governance scales, EDM was able to begin shifting away from a paradigm of reactivity.

The value of such collaborative engagements therefore lies not only in their tangible outcomes (such as the formal adaptation plans that are developed), but also in the informal institutions that arise in the form of social learning and capacity building (Collins and Ison 2009; Emerson et al. 2012), which Armitage et al. (2011) argue are themselves a key form of adaptation. Whilst BRM was the only Municipality to finalise and adopt a climate change adaptation plan in the Province in the first few years and mainstream it into its IDP, it was agreed by officials from all three municipalities that being involved in the process was in itself a highly beneficial and educational experience. In DM, the adaptation planning process gave rise to social learning not only for the municipal stakeholders, but for the provincial CCMSP team as a whole. Specifically, they realised the strong ties between adaptation and mitigation at the local level, and through engagements with local officials from the Municipality, decided that these response programmes should not be separated in the future.

Lesson 3: the importance of mainstreaming climate change into development planning

In South Africa, the mandate to address climate change at the municipal level is generally seen as synonymous with local environmental affairs. In all three municipalities considered in this study, climate change was automatically understood as being a subsidiary function of environmental management. Even in the Drakenstein and Bergrivier Municipalities, in which there is no overarching environmental department, the climate change agenda was assigned to the individual responsible for environmental concerns. The framing of climate change as an environmental issue is problematic in light of the broader developmental context in which the municipalities are positioned, wherein demands for service delivery, economic growth and social upliftment tend to outweigh any concerns related to the environment (Pasquini et al. 2013; Patel 2009). This is particularly true in relation to climate change, which is "still widely perceived within municipalities as a call to respond to a vague and distant threat that complicates (and thereby delays, protracts and makes more costly) immediate tasks of extending public infrastructure and rolling out basic services as quickly, extensively and cheaply as possible to improve living conditions and economic activity in the near term" (Taylor et al. 2014, p. 93).

Many local governments are therefore focussing on the integration of adaptation into existing policies and practices (Anguelovski et al. 2014; Pasquini et al. 2014; Sharma and Tomar 2010). Climate change was acknowledged in all three municipalities as being important, and in some cases fundamental to the way in which development programmes are carried out. However, whilst the majority of interview respondents envisioned a more resilient and climate-considerate future society in theory, it was evident that more immediately pressing issues posed a challenge in terms of prioritising and mainstreaming the climate adaptation agenda into development planning in a practical sense.

The three case studies indicate that the institutional location of climate change plays a significant role in mainstreaming. In comparing the BRM and DM case studies, one can determine that a more beneficial outcome can be achieved if the climate adaptation agenda is steered from a 'higher' (more strategic) institutional standpoint within the municipality. The BRM process, in which an adaptation plan was rapidly finalised and mainstreamed, was championed by the Manager: Strategic Services, who was concurrently responsible for the IDP and worked from within the office of the administrative CEO. Her senior position provided her with a more influential role within the organisation, and she was therefore able to influence the policy process and drive the rapid acceptance of the adaptation plan. Positioning climate change at a more strategic level thus allows it to be understood as cross-cutting (rather than just an environmental) issue. Conversely, in DM, the adaptation planning process was championed from within the Civil Engineering Department by the municipality's senior engineer, who had the co-responsibility of environmental management. This caused confusion around the siting of the adaptation mandate and communication breakdowns across departmental divides, resulting in the adaptation plan being neither completed nor mainstreamed.

In the Eden Municipality, the climate change champion was initially located in the Environmental Department. However, a realisation of the close tie between environmental and disaster management led to the process being unofficially co-championed by the district's disaster manager. The cross-sectoral championing of the climate adaptation agenda proved to be a constructive as it contributed to increased institutional learning around the transversality of the climate change issue.

Conclusion

In conclusion, this chapter has shown that valuable lessons can be garnered from reflecting on 'learning-by-doing' adaptation responses at the municipal level. The CCMSP was a new programme that was experimental in nature, which led to significant variations in the outcomes of each municipality's adaptation planning process. The findings suggest that a number of factors played a role in the degree to which each municipality was successful in producing an adaptation plan. Key in this regard was strong political will and leadership from within the municipality, as well the presence of collaborative partnerships and relationships among stakeholders, including those external to the municipality. In the case of South Africa's institutional context, mainstreaming climate change into master planning documents, particularly the IDP, proved important for getting the required support and resources. To increase the likelihood of mainstreaming, climate change portfolios and champions should be situated in departments other than, or in addition to, environmental departments.

Given the importance of getting multiple actors involved and invested in climate change adaptation, it is clear that co-production was central to both the quality of the process and the product. Integrative leadership skills and the presence of existing relationships and networks proved to be catalysts in this regard. One of the important contributions to increasing the scientific input to the process was partnerships with external institutions. Similarly, when the leading actors were willing and open to learning from and building relationships with other departments, the process was strengthened further. The collaborative processes demonstrated transformative potential in the sense that local community members were able to develop leadership skills as well as form new partnerships across departments and with external stakeholders. Significant learning occurred among both stakeholders and the provincial CCMSP team. This suggests the emergence of more collaborative and learning-centred forms of local adaptation governance. However, in order to gather the diverse and appropriate people, political will and suitable leadership needs to be established, which is often not part of the adaptation mandate. Future adaptation planning processes should therefore place more of an emphasis on tackling the issues of inadequate political will and type of leadership ability, which need to be addressed first and foremost.

Notes

1 Municipal governments are mandated to deliver adequate basic services in an equitable, efficient and sustainable manner, as expressed in the White Paper on Local Government (RSA, 1998) and the Municipal Systems Act (RSA, 2000).

2 SANParks is the body responsible for managing South Africa's 19 national parks, which cover an area of approximately 73,000 hectares. Cape Nature (officially known as the Western Cape Nature Conservation Board) is a provincial government organisation that is responsible for maintaining wilderness areas and public nature reserves in the Western Cape Province.

References

Amundsen, H., Berglund, F., and Westskogh, H. (2010). Overcoming barriers to climate change adaptation-a question of multilevel governance? *Environment and Planning C: Government and Policy*, 28, 276–289. Retrieved from http://doi.org/10.1068/c0941

Anguelovski, I., and Carmin, J. (2011). Something borrowed, everything new: Innovation and institutionalization in urban climate governance. *Current Opinion in Environmental Sustainability*, 3(3), 169–175. Retrieved from http://doi.org/10.1016/j.cosust.2010.12.017

Anguelovski, I., Chu, E., and Carmin, J. (2014). Variations in approaches to urban climate adaptation: Experiences and experimentation from the global South. *Global Environmental Change*, 27(1), 156–167. Retrieved from http://doi.org/10.1016/j.gloenvcha.2014.05.010

Ansell, C., and Gash, A. (2008). Collaborative governance in theory and practice. *Journal of Public Administration Research and Theory*, 18(4), 543–571. Retrieved from http://doi.org/10.1093/jopart/mum032

Armitage, D., Berkes, F., Dale, A., Kocho-Schellenberg, E., and Patton, E. (2011). Co-management and the co-production of knowledge: Learning to adapt in Canada's Arctic. *Global Environmental Change*, 21(3), 995–1004. Retrieved from http://doi.org/10.1016/j.gloenvcha.2011.04.006

Barton, J.R., Krellenberg, K., Harris, J.M., Richard, J., Krellenberg, K., and Harris, J.M. (2015). Collaborative governance and the challenges of participatory climate change adaptation planning in Santiago de Chile. *Climate and Development*, 7(2), 175–184. Retrieved from http://doi.org/10.1080/17565529.2014.934773

Bergrivier Municipality (BRM). (2014). *Bergrivier Municipality Climate Change Adaptation Plan*. Piketberg: Bergrivier Municipality.

Bulkeley, H., and Castan Broto, V. (2013). Government by experiment? Global cities and the governing of climate change. *Transactions of the Institute of British Geographers*, 38(3), 361–375. Retrieved from http://doi.org/10.1111/j.1475-5661.2012.00535.x

Carmin, J., Anguelovski, I., and Roberts, D. (2012). Urban climate adaptation in the global South: Planning in an emerging policy Domain. *Journal of Planning Education and Research*, 32(1), 18–32. Retrieved from http://doi.org/10.1177/0739456X11430951

Carmin, J., Roberts, D., and Anguelovski, I. (2009). Planning climate resilient cities: Early lessons from early adapters. *World Bank 5th Urban Research Symposium, Cities and Climate Change*, June, 30. Retrieved from http://siteresources.worldbank.org/INTURBANDEVELOPMENT/Resources/336387-1256566800920/6505269-1268260567624/Carmin.pdf

Chu, E. (2015). The political economy of urban climate adaptation and development planning in Surat, India. *Environment and Planning C: Politics and Space*, 34(2), 281–298. Retrieved from http://doi.org/10.1177/0263774X15614174

Chu, E., Anguelovski, I., and Carmin, J. (2015). Inclusive approaches to urban climate adaptation planning and implementation in the Global South. *Climate Policy*, May, 3062, 1–21. Retrieved from http://doi.org/10.1080/14693062.2015.1019822

Clarke, B., Stocker, L., Coffey, B., Leith, P., Harvey, N., Baldwin, C., . . . Cannard, T. (2013). Enhancing the knowledge-governance interface: Coasts, climate and collaboration. *Ocean and Coastal Management*, 86, 88–99. Retrieved from http://doi.org/10.1016/j.ocecoaman (2013.02.09).

Clifford-Holmes, J.K., Slinger, J.H., Mbulawa, P., and Palmer, C.G. (2015). Modes of failure of South African local government in the water services sector. *Conference Proceedings of the 33rd International Conference of the System Dynamics Society*, 19–23 July, 1–17.

Cloutier, G., Joerin, F., Dubois, C., Labarthe, M., Legay, C., and Viens, D. (2014). Planning adaptation based on local actors' knowledge and participation: A climate governance experiment. *Climate Policy*, January, 14, xx. Retrieved from http://doi.org/10.1080/14693062.2014.937388

Collins, K., and Ison, R. (2009). Jumping off Arnstein's ladder: Social learning as a new policy paradigm for climate change adaptation. *Environmental Policy and Governance*, 19(6), 358–373. Retrieved from http://doi.org/10.1002/eet.523

Crosby, B.C., and Bryson, J.M. (2010). Integrative leadership and the creation and maintenance of cross-sector collaborations. *Leadership Quarterly*, 21(2), 211–230. Retrieved from http://doi.org/10.1016/j.leaqua.2010.01.003

Emerson, K., Nabatchi, T., and Balogh, S. (2012). An integrative framework for collaborative governance. *Journal of Public Administration Research and Theory*, 22(1), 1–29. Retrieved from http://doi.org/10.1093/jopart/mur011

Henstra, D. (2015). The tools of climate adaptation policy: Analysing instruments and instrument selection. *Climate Policy*, April, 3062, 1–26. Retrieved from http://doi.org/10.1080/14693062.2015.1015946

Leck, H., and Simon, D. (2013). Fostering multiscalar collaboration and co-operation for effective governance of climate change adaptation. *Urban Studies*, 50(6), 1221–1238. Retrieved from http://doi.org/10.1177/0042098012461675

Mees, H.L.P., Driessen, P.P.J., Runhaar, H.A.C., and Stamatelos, J. (2015). Who governs climate adaptation? Getting green roofs for stormwater retention off the ground. *Journal of Environmental Planning and Management*, 56(6), 802–825. Retrieved from http://doi.org/10.1080/09640568.2012.706600

Nel, J.L., Le Maitre, D.C., Nel, D.C., Reyers, B., Archibald, S., Van Wilgen, B.W., . . . Barwell, L. (2014). Natural hazards in a changing world: A case for ecosystem-based management. *PLoS ONE*, 9(5). Retrieved from http://doi.org/10.1371/journal.pone.0095942

Pasquini, L., Cowling, R.M., and Ziervogel, G. (2013). Facing the heat: Barriers to mainstreaming climate change adaptation in local government in the Western Cape Province, South Africa. *Habitat International*, July, 40, 225–232. Retrieved from http://doi.org/10.1016/j.habitatint.2013.05.003

Pasquini, L., and Shearing, C. (2014). Municipalities, politics, and climate change: An example of the process of institutionalizing an environmental Agenda within local government. *The Journal of Environment & Development*, 23(2), 271–296. Retrieved from http://doi.org/10.1177/1070496514525406

Pasquini, L., Ziervogel, G., Cowling, R.M., and Shearing, C. (2014). What enables local governments to mainstream climate change adaptation? Lessons learned from two municipal case studies in the Western Cape, South Africa. *Climate and Development*, March, 7, 60–70. Retrieved from http://doi.org/10.1080/17565529.2014.886994

Patel, Z. (2009). Environmental justice in South Africa: tools and trade-offs. *Social Dynamics*, 35(1), 94–110. Retrieved from http://doi.org/10.1080/02533950802666956

Republic of South Africa (RSA) (1998a). *Local Government: Municipal Structures Act* (Act no. 117 of 1998).

Republic of South Africa (RSA) (1998b). *White Paper on Local Government*. Pretoria.

Republic of South Africa (RSA). (2000). *Municipal Systems Act* (Act No. 32 of 2000). Government Gazette no. 21776. Pretoria.

Roberts, D. (2008). Thinking globally, acting locally – Institutionalizing climate change at the local government level in Durban, South Africa. *Environment and Urbanization*, 20(2), 521–537. Retrieved from http://doi.org/10.1177/0956247808096126

Roberts, D., Boon, R., Diederichs, N., Douwes, E., Govender, N., Mcinnes, A., . . . Spires, M. (2012). Exploring ecosystem-based adaptation in Durban, South Africa: "Learning-by-doing" at the local government coal face. *Environment and Urbanization*, 24(1), 167–195. Retrieved from http://doi.org/10.1177/0956247811431412

Roberts, D., and O'Donoghue, S. (2012). Urban environmental challenges and climate change action in Durban, South Africa. *Environment and Urbanization*, 25(2), 1–21. Retrieved from http://doi.org/10.1177/0956247812456472

Sharma, D., and Tomar, S. (2010). Mainstreaming climate change adaptation in Indian cities. *Environment and Urbanization*, 22(2), 451–465. Retrieved from http://doi.org/10.1177/0956247810377390

Snowden, D.J., and Boone, M.E. (2007). A leader's framework for decision making a leader's framework for decision making. *Harvard Business Review*, 85(11), 68–76, 149. Retrieved from http://doi.org/10.1109/MCDM.2007.369449

Spires, M.H. (2015). *Barriers to and Enablers of Climate Change Adaptation in Four South African Municipalities, and Implications for Community Based Adaptation*. Department of Environmental Science, South Africa: Rhodes University.

Swilling, M. (2014). Rethinking the science – Policy interface in South Africa : Experiments in knowledge co-production. *South African Journal of Science*, 110(5), 1–7.

Taylor, A., Cartwright, A., and Sutherland, C. (2014). Institutional pathways for local climate adaptation : A comparison of three South African municipalities, March, 140.

Turner, S., Fünfgeld, H., and Robertson, S. (2016). Strategies for embedding climate change adaptation in public sector organisations. Melbourne: Centre for Urban Research. Retrieved from http://doi.org/10.13140/RG.2.1.1200.7449

Ward, K. (2010). Towards a relational comparative approach to the study of cities. *Progress in Human Geography*, 34(4), 471–487. Retrieved from http://doi.org/10.1177/0309132509350239

WCG (Western Cape Government) (2014a). *Socio-Economic Profile Eden District Working paper*. Cape Town: Western Cape Provincial Treasury.

WCG (Western Cape Government) (2014b). *Western Cape Climate Change Response Strategy*. Cape Town: Department of Environmental Affairs and Development Planning (DEA & DP).

Ziervogel, G., New, M., Archer Van Garderen, E., Midgley, G., Taylor, A., Hamann, R., . . . Warburton, M. (2014). Climate change impacts and adaptation in South Africa. *WIREs Clim Change*, October, 5, 605–620. Retrieved from http://doi.org/10.1002/wcc.295

Ziervogel, G., and Taylor, A. (2008). Feeling stressed: Integrating climate adaptation with other priorities in South Africa. *Environment Science and Policy for Sustainable Development*, 50(2), 32–41. Retrieved from http://doi.org/10.1038/vital795

5 Towns in transition – regional and ideological diversity among local climate protection projects and regional revitalisation efforts in rural Japan

Christian Dimmer and Daniel Kremers

Introduction

Japan today faces two major socio-economic challenges: fast demographic ageing and a high level of energy dependency and insecurity. Both topics are closely linked to the livelihood and prosperity of local communities and climate change mitigation (cf DeWit and Tani 2008). Demographic ageing comes with urbanisation, and while the population in the urban centres grows, many local communities and regions are in decline. Meanwhile rural communities are an essential asset for Japan's self-reliance in terms of food, resources and energy – and not just because all of Japan's atomic power plants were placed in the countryside, far away from the cities where most of the energy is consumed. The Great East Japan Earthquake of March 11, 2011, and the ensuing accident at the Fukushima Dai'ichi nuclear plant, however, not only led to the temporary shutdown of all nuclear reactors, but the unprecedented energy crisis also catapulted renewable energy from the margins into the centre of policy making and everyday discourses in Japan.

Although Japan's government has been promoting renewable energies through a feed-in tariff system since 2012, the fear of crippling power outages led energy suppliers to ramp up imports of fossil fuels – initially liquified natural gas (LNG) and then increasingly the far cheaper coal. Paradoxically, one year after the COP21 in Paris, where 196 nations pledged to keep global warming below the threshold of two degrees Celsius, the situation in Japan is mixed. "Power demand is weak, solar capacity is increasing at breakneck speeds, nuclear capacity is returning, and coal-fired generation is rising" (Tsukimori and Sheldrick 2016). At the same time, renewable energy producers are facing great difficulties because the existing power grids are not able to handle the additional, often fluctuating supply of wind and solar energy that cause grid stability issues.

While the Japanese central government aims at reducing its reliance on fossil fuels in the long run by retaining a comparatively high share of climate-neutral nuclear power, many local initiatives seem to have understood the urgency of the situation and show more concern for global and local environmental issues (METI 2014, p. 21). Often, however, these projects are not merely framed from an explicitly environmental perspective or by taking recourse to a global, highly moralised narrative of planetary environmental crisis, but more to practical, locally based narratives, closer to the everyday life worlds, and strongly concerned with Japan's rural population decline, and hyper-ageing.

While Japan is one of the most urbanised countries in the world, with 77 percent of the population concentrating in 76 functional urban areas, and 23 percent in small and mid-level municipalities (OECD 2016, p. 133), the role of the country's regions are crucial for the safeguarding of an independent food, water, energy and resource supply, as well maintaining biodiversity. While many climate change models assume that when population declines, gains in fields such as energy consumption, carbon output, and biodiversity are naturally to be had, recent research suggests that CO_2 emissions can increase when populations thin out and more elderly people live alone, distributed over a wide territory (Bird 2014; Matanle 2016). There is therefore a significant link between climate action and regional revitalisation policies. According to projections from the National Institute of Population and Social Security Research (NIPSSR), the country's population will decline by somewhere between 26 percent and 38 percent by 2060, and around 40 percent of people will be over age 65 (Bird 2014).

In this chapter we are looking at two cases of local climate action in peripheral Japan, in regions outside of the country's metropolitan core areas: 1) Town A, which is located on the southernmost main island Kyûshû and 2) District b of City B, a formerly independent mountain hamlet in central Japan that was recently merged with seven other communities, into a now massively enlarged city. With these two case studies, we demonstrate that local climate initiatives within Japan differ from national agendas, but also from each other. We show that distinct geographic, demographic and political conditions can make a critical difference when it comes to building and implementing sustainability strategies, and we point towards factors that support or hinder the implementation and mainstreaming of climate change action. Although far from exhaustive, our analysis shows that climate change–related activities in Japan are often not primarily addressed or framed with climate change as the main issue. Our cases indicate that local problems such as depopulation, lack of economic opportunities or hyper-ageing figure equally prominent in the perception and motivation for taking action, with climate protection often as a side effect. We also point out that while there is a huge natural, economic and social potential for renewable energies in Japan, committed politicians, entrepreneurs and citizen activists still face serious political and economic challenges, despite climate protection now figuring prominently on the national agenda.

This chapter is intended as a call to develop a more nuanced understanding of the differing and contingent ways in which climate action is framed locally in Japan and elsewhere, and how the diverse contingent social, historical and geographic factors contribute to the specific climate change policy response. In order to achieve this, we will first situate our two case studies in their respective spatial context, showing, for example, how geographic features of the cultural landscape and local climate, or their accessibility from and relations with nearby urban areas, influence the interplay between endogenous potentials and external influences. Furthermore, we provide a discussion of the country's governance structure and recent structural changes in order to highlight critical institutional challenges that local stakeholders are struggling with. We will highlight the types of local actions and collaborations that are emerging, their characteristics and outcomes and what can be learned from these to inform future local actions elsewhere. The research is based on field visits in autumn 2015, during which extensive interviews with local citizens and leaders were carried out.

National frameworks, regional disparities and local conditions

To understand the challenges and opportunities local communities and activism in Japan face today, we have to take into account geographic characteristics – including topography and natural resources – the level of administrative independence and political mobilisation as well as the current trend of demographic ageing, paired with urbanisation and regional decline. Furthermore, to appreciate the contribution to climate mitigation by local actors, Japan's national climate policy needs to be understood.

Japan's climate policy goals

Reducing environmental pollution and conserving energy has been a major concern for industrial and environmental policies in Japan since the 1970s (Stijn 2014, p. 17). Throughout the 1950s and '60s, thousands of environmental protests sprang up across Japan, against the massive environmental degradation and human suffering that resulted from the government's single-minded, pro-growth industrial policies. In famous pollution cases like the Minamata mercury poisoning, the government sided with the polluters to maintain the status quo that favoured the profits of big a corporation at the expense of the quality of life of residents (Osiander 2007). It would even "shut down several university research projects into the causes of pollution diseases, [while] the companies obstructed efforts to determine the causes and nature of the pollution problems" (Sorensen 2002, p. 204). These protests translated into a growing electoral crisis of the Liberal Democratic Party (LDP) and brought many progressive local politicians into office, who promised to solve the environmental crises (MacDougall 1980, p. 84). Though modern Japan (since 1868) has a history of protest and civil movements, these appeared spatially, temporarily, thematically disconnected and were never sustained and strong enough to influence the state and its institutions. In Japan these rather short moments of public dissent are not publicly recognised as moments of social progress but as signs of crisis and unrest (cf. Avenell 2010; cf Andrews 2016). Some argue, however, that local environmental movements have played a crucial role in the maturing of civil society in Japan (cf Yamamoto 1999; Yoshida 1999).

The impact of the oil shocks of 1973 and 1979 provided another strong incentive for the Japanese economy to reduce its dependency on mineral oil imports. In 1974 the government launched the so-called Sunshine Project to promote solar and geothermal energy generation, and between 1979 and 2009 energy efficiency had been improved by roughly 33 percent (Stijn 2014, p. 18). With CO_2 reduction becoming more important in the light of global warming and climate protection from the late 1990s on, the Japanese government sought to increase its reliance on emission-free nuclear power. The 2010 Basic Energy Plan issued by the Ministry of Economy, Trade, and Industry (METI) envisioned an increase in electric power generated through nuclear energy from 30 to 50 percent by 2030 (Huenteler et al. 2012, p. 6). However, the Great East Japan Earthquake of March 11, 2011 (hereafter 3.11), and the ensuing energy crisis that resulted from the shutdown of all nuclear plants in Japan, as a consequence of the nuclear crisis at the Fukushima Dai'ichi power plant, has put these plans for a nuclear future into jeopardy.

For the United Nations Climate Change Conference of Parties (COP21/CMP11) in Paris, the government of Japan announced to reduce greenhouse gas (GHG) emissions "through [. . .] leading technologies and support for developing countries", stating that "due to the Great East Japan Earthquake and the accident at the [. . .] Fukushima Dai-ichi Nuclear Power Station" the country had to rebuild its "energy [and climate] strategy from scratch" (GOJ 2015, Internet). After the meltdown of three out of six reactors in the tsunami-stricken nuclear power plant Fukushima 1, Japan's nuclear power plants one after another were taken offline for inspection and stress tests, and as of the end of 2016, only one has been returned to operational status.

On the one hand, the government-induced culture of energy saving (*setsuden no bunka*) quickly spread throughout society and helped to lower the energy demand significantly (cf. Lindner 2014). On the other hand, the resulting drop in electric power capacity was quickly offset by an increased import of fossil LNG to be combusted in power plants. This, however, prompted a significant increase in emissions of CO_2 and other GHG. As a result, the planned reduction of emissions of 26 percent from 2013 to 2030, amount to only 25.4 percent, if 2005 is selected as the base year. The government predicted Japan's annual emissions for 2030 as 1.042 billion t-CO_2 equivalent (ibid.). These figures are, however, based on the assumption that Japan will be able to put into practice its "energy mix", containing at least 20 percent of electricity generated in nuclear power plants as laid out in the METI's "Long-term Energy Supply and Demand Outlook" of July 16, 2015 (ibid.).

A feed-in tariff system (FIT) to promote renewable energies was implemented by a Democratic Party of Japan (DPJ) headed reform government, led by Naoto Kan in 2012 (see table 5.1). At this moment mostly solar energy is produced, which accounts for just under 97 percent (27.4 GW) of the 28.4 GW of renewable capacity as of June 2016 (DeWitt 2016, p. 10). Most of these facilities are large-scale and run by big corporations. However, the utilities can refuse to feed in renewable energies from local suppliers on a case-by-case basis, when they

Table 5.1 Feed-in tariffs from 2012 to 2015 according to energy source and technology

Source	*Technology/size*	*FIT 2012*	*FIT 2013*	*FIT 2014*	*FIT 2015*	*Purchase period*
		Yen/kWh				*Years*
Solar	PV (<10 kW)	42	38	37	33	10
	PV (≥10 kW)	40	36	32	27	
	Double generation (<10 kW)	34	31	31	27	
Wind	On shore (<20 kW)	55			55	20
	On shore (≥20 kW)	22			22	
Geo-thermal	<15 MW	40	40	40	40	15
	≥15 MW	26	26	26	26	
Hydro power (new facilities)	<200 kW	34	34	34	34	20
	200–1000 kW	29	29	29	29	
	1–30 MW	24	24	24	24	
Biogas		39	39	39	39	20
Dry Biomass	Forest <2 MW	32	32	32	40	20
	thinning ≥2 MW	32	32	32	32	
	Wood, crop residues	24	24	24	24	
	Wood (construction debris)	13	13	13	13	
	Waste incineration	17	17	17	17	

Source: METI (2015, Internet)

claim that the power grid destabilised. So far the grid, the main power plants, as well as the supply are owned and operated by nine regional monopolies. This makes it difficult for new suppliers to gain access to the market. Unbundling the business of grid and supply is scheduled for 2020.

While the generous FIT has sparked a hike in the installations of photovoltaic (PV) in 2012, the installation and utilisation of renewable energies differ significantly from place to place. The effects of national policies thus have to be evaluated on a case-by-case basis that takes into account regional and local geographical, political and socio-economic factors. In the following case studies, we will comparatively analyse different approaches and speeds of energy transition and climate change mitigation on the local level.

Geographic and municipal organisation

The Japanese archipelago is part of the Pacific Ring of Fire and seated on the edges of the Asiatic, the Philippine, the North American and the Pacific tectonic plates. Thus Japan is prone to natural disasters such as earthquakes, volcano eruptions, tsunamis and typhoons, which is why resilience thinking plays a key role in recent plans and policies (cf OECD 2016, 79) but is also reflected in local building traditions. The high level of seismic activity has also produced a vast mountainous terrain with narrow deep valleys, large forested areas and comparably little space suitable for settlement and agriculture. Thus today's agriculture, industry and urban settlements compete for relatively limited space in alluvial plains on the east coast of Honshû (Flüchter 1998, p. 16).

Together with neighbouring prefectures the wider Tokyo Metropolitan region is inhabited by almost 46.9 million people, followed by the second largest agglomeration around Osaka, Kobe and Kyoto with around 17 million inhabitants, and the region around Nagoya with 11.4 million inhabitants. These three densely populated agglomerations together host ca. 75 million inhabitants – half of Japan's total population – while the rest of the country is significantly less densely populated. Despite various attempts of decentralisation over the last decades, Japan's metropolitan cores continue to grow at the expense of the rural parts of the country.

In order to deal with above-mentioned unbalanced national development of rapidly growing metropolitan areas and depopulating rural regions, the government repeatedly encouraged municipal mergers (*gappei*), to create larger municipalities that it saw better capable of providing public services more cost-efficiently (Amai 2010). Thus, the number of villages, towns and cities has been drastically reduced over the last century and many "rural municipalities [. . .] have, in effect, disappeared" from the map (Rausch 2008, p. 225). The last major phase of municipal mergers lasted from 1999 till 2010 (*heisei gappei*), through which the number of local municipalities was reduced to 1.741 as of in March 2010. Although these amalgamations were accompanied by considerable public debate, it seems fair to say that economies of scale and efficiency considerations were constantly valued higher than local identity and civic pride or forms of governance that are closer to the residents and more easily accessible (cf Rausch 2006; Reiher 2008, p. 166). What Samuels called "supersizing" (Samuels 2013, p. 175) would lead to fewer public facilities, fewer government officials, less democratic representation and a loss of local autonomy and initiative of hitherto independent communities.

Our case studies reflect this regional diversity of overall Japan: Town A is located in a flood-prone alluvial plain on the southern main island of Kyûshû, well-connected to higher-tier infrastructure and proximate to the island's main city Fukuoka. It is an independent municipality because its citizens and the local government have resisted repeated attempts of mergers with its surrounding municipalities with their civic pride and resources strong enough to remain independent. District b of City B is a remote mountain village in central Japan that – despite its remoteness – plays a significant role for regional energy and resource supply. It hosts a hydroelectric dam, is a vital water protection area supplying water to the wider region, is a regional centre for winter sports and is a natural reserve. It has only recently lost its administrative independence when it was merged with other municipalities into City B.

Governance system and local self-administration

Before a closer analysis of our case studies, we need to highlight the relatively weak role of local governments within Japan's overall highly centralised governance system. Different administrative practices such as frequent job rotations inside city halls, or lacking autonomy vis-a-vis higher-tier governments can partly explain the difficulties for Japanese local governments taking a stronger local initiative.

While the central government promoted municipal mergers, it also undertook measures of decentralisation, transferring many administrative tasks from the national and regional level to the municipalities. It is instructive that a policy field as crucial for local development, such as urban planning, was outside the control of all but the largest municipalities up until 1999, when finally, more autonomy was granted. Thus, there is only a very short institutional history of local initiative and planning expertise and personnel capacities and local skills are in short supply. Thus, the central government delegated more administrative functions to the municipalities but withheld financial autonomy, which stymied local autonomy and initiative (Samuels 2013, p. 152).

Meanwhile, demographic and economic disparities between the agglomerations on Japan's South-Eastern Pacific coast and the rural peripheries on the west coast have increased dramatically. The significant concentration of population and productivity in the metropolitan agglomerations stands in sharp contrast to an unbroken trend of emptying out and obliteration of the countryside and few rural local governments have the capacities to cope with this challenge. While Japan's post-war developmental welfare state model provided little redistribution among social classes, i.e., from rich to poor, financial transfers from the urban centres to the peripheries and subsidies for infrastructure projects have been very generous (Chiavacci and Hommerich 2016, p. 19). The reason for this is political. It has been argued, that through subsidies for developing local infrastructure, the ruling Liberal Democratic Party (LDP) was able to generate and sustain a strong voter support in the rural areas of Japan (Sorensen 2002, p. 208). Due to two decades of economic stagnation and a skyrocketing national deficit, the national government was forced to reduce such redistributive measures and instead encourages now open competition between rural communities. Meanwhile local communities are the most affected by Japan's demographic ageing and shrinking.

These geographical conditions and socio-economic and political developments are reflected in the local communities we studied. Town A is located in a prefecture on the southern island of Kyûshû that features both heavy industries as well as agriculture. Its municipal independence combined with strong civic capacities allowed the local authorities to develop and implement a far-reaching sustainability strategy. In contrast, after a recent merger with seven formerly independent neighbouring cities, towns and villages in 2005, the formerly independent mountain village (*son*) b became a remote, outlying district (*chiku*) of a newly formed City B. Compared to Town A, District b not only lacks political autonomy, but is further disadvantaged by its remote location, far away from the city hall and other public facilities. While renewable energies can help to create alternative economic opportunities to narrow the growing urban-rural inequality, the inaccessibility of the political process and the lack of governance resources frustrate existing sustainably initiatives and also suffocate local initiatives.

Demographic development, austerity, and regional revitalisation policies

In order to assess the opportunities and constraints of actors in local climate action, not only direct climate policies and plans are relevant, but also broader regional revitalisation policies – a policy field that is rapidly gaining momentum in recent years. The declining birth rate (*shoshika*) and the rapid ageing (*koreika*) of Japan's population have been major political concerns since the mid-2000s. Already in as early as 1967 the term *depopulating regions* (*kaso chiiki*) had been coined, and in 1970 the Japanese government launched the *Urgent Measure Act against Depopulation of Regions* (*kaso chiiki taisaku kinkyû sochihô*) (Elis and Lützeler 2008, p. 17). Despite decades of revitalisation efforts, rural regions continue to depopulate. "Soft," community development was often neglected, while subsidies for construction and public works created an excessive supply of "hard" infrastructure such as roads, public buildings, business parks, sport arenas or dams that created maintenance burdens for future generations, caused a high dependency on the construction sector and further aggravated spatial disparities (Feldhoff 2008, p. 39).

While the issue was outshone by an aggressive focus on geopolitics, monetary easing and controversial security laws under the second government under Shinzō Abe (2012–), recently, the national government identified regional development as key issue and launched several government bodies and initiatives under the slogan "Creation of Regions" (*chihô sôsei*). The neologism *sôsei* combines the words *saisei* (revival/ renewal) and *sôsei* (creation/ genesis) and is difficult to translate. This move was partially prompted by the 2014 best-selling book "Local Extinction – The over-concentration on Tokyo brings about a rapid population decline" (*Chihō Shōmetsu – Tokyo Ikkyoku Shūchū ga maneku Jinkō Kyūgen*) by Hiroya Masuda. Masuda, former governor of the northeastern prefecture of Iwate and head of a government committee on local revitalisation, links Japan's regional disparities to the concentration of economic and political functions in Tokyo. Based on the latest official figures from the government's NIPSSR, he predicts that no less than 896 cities, towns and villages throughout Japan out of a total of 1,737 will face extinction by 2040.

In September 2014 Prime Minister Abe appointed Shigeru Ishiba, a former Minister of Defence and promoter of nuclear energy and rearmament (Dawson 2011), as cabinet minister for regional economic revitalisation with sole objective to promote the revitalisation of Japan's depopulating and rapidly ageing rural towns and villages. How important the issue is, is shown by the fact that two high-level bodies were created – the *Bureau for the Promotion of Regional Creation* at the Cabinet Secretariat and the *Headquarter for the Creation of Towns, Humans, and Jobs* at the Cabinet Office (Kantei 2016a, 2016b). The government's public relations office also runs a website called "*Ima koso chihô sôsei* (Now is the time for the Creation of Regions)" (Gov-online 2016). In the same year the government also announced to spent 370 billion yen on regional development between 2014 and 2016 (Kantei 2016b).

Nonetheless, most of these initiatives are strongly growth oriented and follow a neoliberal agenda. Ishiba, replaced by Kozo Yamamoto in August 2016 (NKS 2016.08.04), stressed for example that "If we equalise things among those [municipalities] who make an effort and those who don't, the whole nation will collapse . . . Each village, town and city needs [to compete and] work out how to revive their community" (Bloomberg 2015). Since 2014, the subject of regional revitalisation is widely discussed not only among policy and planning experts but also in the wider public sphere and in mainstream media. The mid- to long-term effects of these policies, however, have yet to be seen.

Case studies – interplay of local political cultures with national guidelines

While national governments have made far-reaching climate protection pledges at the COP21, these national targets will ultimately need to be translated into new, stricter national legislation, far-reaching new energy policies, but also broken down and implemented at the local level. Thus local governments across the planet are facing the challenge to devise climate protection plans and to implement concrete projects, while at the same time being confronted with similar issues of depopulation, out-migration and socio-economic change. We present two case studies from small rural communities in Japan in order to demonstrate if and how government goals *currently* translate into local climate action. One is a nationwide recognised environmental model town (Town A), while the other (District b) represents the average peripheral community that has some potential but lacks governance capacities.

Town A is located in the southwest of Japan's southern main island Kyûshû and formed in the mid-1950s through a merger of three villages. The only 18km^2 small community has 15,000 inhabitants, and more than half of its territory is covered with rice paddies. Town A boasts very good access, as it is only 60 km away from the regional capital Fukuoka. The next highway intersection is only 20 minutes away, and the closest Japan Railway (JR) station 15 minutes, with access to the bullet train network within 30 minutes. Also a regional private railway stops in town.

Owed to this good access, the population stabilised at the level of the 1950s despite some minor decline during the 1970s. Similar to other places in Japan, the number of households doubled from about 2,500 to 5,000 since the 1950s with an average of three persons. Twenty-one percent of the population is below the age of 20, and 17 percent are older than 70. Roughly 50 percent of the population are in employment. Being in commuting distance to Fukuoka, the service sector makes up 58 percent of all jobs, with 28 percent in construction and manufacturing and a high 14 percent in agriculture. In 2010 there were 45 manufacturing companies with more than three employees, half of them in wood processing and furniture production. Since the 1950s the number of agricultural households is continuing to decline, which parallels developments on a national level. Their share sank from 55 to 20 percent in 2010. While 70 percent of agricultural households generate the majority of their income outside of agriculture, the share of full-time farmers is at 16 percent.

Since a municipal merger with a neighbouring city, two towns, and four other villages in 2005, the formerly independent Village b has become a mere administrative district (District b) of the newly formed City B. It lies in the mountains not far away from the Sea of Japan, at the west coast of Japan's central main island Honshû. The new municipality has a population of roughly 110,000 inhabitants that populate a vast territory of 775 km^2. However, while almost one third of the new, magnified city area is made up of District b, only a mere 1 percent of the city's overall population lives here. While the huge majority of its territory comprises mountainous regions, only a small part of City B is located in the alluvial plains along the coast, where the majority of the growing population lives and where the new city hall is sited. With three railway stations on a JR main line and a bullet train stop only 30 minutes away in the prefectural capital, the residents in the lowlands enjoy favourable access to public transport. District b, however, is located in a distance of 43 km from the city hall and is not connected to the railway network. It takes one hour by car over winding mountain roads to get to the central settlement of District b. The closest railway station is 35 km away. The mountains with heavy snowfalls of up to three meters contribute to its isolation.

The differences in topography, urbanization and transportation within City B result in a stark contrast of population development. While the population in the more urban, coastal districts grew by 9 percent between 1990 and 2010, the very rural mountain districts that constitute 85 percent of the city's territory have experienced a steep 20 percent decline. Due to the mobility of young people and the lack of job alternatives, rural areas like District b also display a faster ageing of their population. Both the share of the working age population and that of residents below the age of 15 is smaller in District b than among the city's total population (tab. 2).

City B also presents recognizable discrepancies in the workforce according to industrial sectors. While services account for more than 60 percent both in the overall and the district's employment rate, the share of jobs in agriculture and forestry in District b is almost three times of that (tab. 3).

Many service jobs of District b are in the tourism sector, with the mountainous region attracting 2.2 million visitors each year. While the number of tourists visiting all mountainous districts has declined by 14 percent overall in recent years, only District b experienced a rise in visitors by 15 percent. The main tourist

Table 5.2 Share of age groups in City B and District b

Age group	*City B*	*District b*
	Percentage of population	
Younger than 15	15	11.5
15–64	64	53
65 and older	21	35

Table 5.3 Share of jobs by industry in City B and District b

Industrial sector	*City B*	*District b*
	Percentage of employment	
Agriculture and forestry (1st)	3.4	9
Manufacturing and construction (2nd)	21.7	29.5
Services (3rd)	63.9	62

attractions are winter sports and hot springs (*onsen*). The local alpine ski resort is operated by a local non-profit organisation (NPO). While one hot spring hotel has been recently bought by a young local man and is run as a private business, the other hot spring is operated by an NPO called *District b Community Development Council* and functions also as a public meeting place for the mostly elderly residents, next to the central village square.

In order to strengthen visitor numbers, City B succeeded to get large parts of the mountainous area listed as one of Japan's 27 national designated Geo-Parks. The educational concept of the park stresses a combination of geo-diversity, bio-diversity and local culture in this alpine high-altitude environment with its steep rivers and deep gorges. The Geo-Park features 45 exhibits that are scattered over the territory of City B through which the local government is trying to tap into as local income sources and to promote sustainable regional development. However, few residents are aware of this, and few visitors visit the area solely because of the park designation.

The urbanised areas of City B as well as parts of the wider region not only depend on the mountainous hinterland for generating revenues from tourism but also for fresh water supply. In order to protect the drinking water and avoid landslides, the mountain forests need to be retained and nourished. This in turn requires stabilising the residential population and a revitalisation of the forestry that has been declining in this peripheral area since the 1960s. Since local wood is also more expensive than imported lumber, new business innovations are needed.

Town A – environmental protection and climate mitigation through local resources management and circular economy

Town A has become a nation-wide renowned, rather atypical example of a comprehensive, top-down and middle-out strategy that was initiated by a progressive mayor and implemented in a community that has preserved a very strong civic pride. Environmental protection constitutes an essential part of the municipal policies here. They especially stress a reduction of waste and climate protection through the creation of a local circular economy. In 2011, the breakdown of infrastructure and supply systems in the days after the Great East Japan Earthquake highlighted the vulnerability of centralised food and energy supply systems and gave a nationwide boost to the ideas of distributed infrastructure, and more resilient local systems (cf Dimmer 2017).

As motivation for local residents and for communal branding, the numerous climate and environment-related projects of the town have been labelled with an onomatopoetic expression for rolling or circulation, that can be translated as *Curly*. A special mascot (*yure kyara*) – a bee-like character with a green lock on its head holding a magic wand – was specially designed for this purpose. Among the projects under the roof of *Curly* are a biogas plant, a rest stop at a major thoroughfare, a waste collection station, a management and information centre, a public park with sculptures welded from scrap metal and a petting zoo. While the municipal office for circular economy supervises these projects, some are also established as cooperatives and associations. The town's effort was officially acknowledged by the Japanese Ministry of Agriculture, Forestry and Fisheries (MAFF) by acknowledging it as "Biomass-Town".

The heart of the strategy that was initiated by the town's mayor is the *Circulation Centre Curly*. It consists of a biogas plant and an information centre. The town's biogas plant is supplied with wet biomass from households and agriculture, mostly rice husks from rice farming, sludge from pig breeding and food rests from private households and gastronomy. It can process 3.8 tons of organic household waste, 7,000 litres of excrement and 30,600 litres of slurry tank sludge per day. Through anaerobic fermentation, it produces methane and a dry residue. The methane fuels two modified diesel engines that each generate 30 kW electric capacity while the residue is distributed to local farmers as fertilizer. On the same property as the biogas plant is a public building, containing office space and seminar rooms.

Next to the *Circulation Centre*, along a major thoroughfare, is a rest stop called *Curly Rest Stop Town* A that is operated by an association named *Sustainable Town* A. It hosts a gastronomic service called *Deli & Buffet Curly* that serves and promotes locally produced food and other products. It has been well-received by travellers and locals alike and has developed into an important meeting spot for the community. The building also hosts a shop for locally produced groceries called *Curly Dream Market*, which is operated by the local section of the prefectural JA (Central Union of Agricultural Co-operatives) farming cooperative. Outside the rest stop is a playground and a little enclosure with goats and rabbits referred to as *Everybody's Open Space to Rest Curly Park*, which offers playful environmental education for children and parents.

Near the rest stop, the municipality operates a collection station for waste to be separated for recycling called *Town A Environment Plaza*. There are roughly 20 different categories of waste to be separated and collected here. Local citizens bring in their waste themselves and as an incentive, are exempted from fees for public waste management. Next to the station there is a small building which hosts a used goods shop called *Re-Use Plaza* "*Gogo*".

Environmental and climate protection has been high on the political agenda of Town A. In March 2008, the municipal assembly issued a "zero waste declaration" in which the town declared the ambitious aim to completely stop burning or land-filling waste by 2016. Though the declaration's central point is waste reduction, it is embedded in a wider discourse on generational justice and climate protection. Its preamble begins with the statement that "the future of children is in danger" and that "global warming" and "climate change" were "threatening the existence of the human species". The preamble is also unambiguous about the fact that the "cause for this are human activities and a society based on the large scale consumption of resources". The municipality therefore aimed at "correcting" this "very wasteful lifestyle" and creating a "sustainable town" that does not live on the expense of future generations. It makes further reference to the 2006 documentary movie *An Inconvenient Truth* and the 2007 IPCC's projection of a rise in global average temperature of 6.4 degrees by 2100.

As concrete measures the declaration announces 1) to foster an education for a "thrifty heart" in children and adults, and 2) practices that "reduce, reuse, recycle" (3R) materials and resources. It further 3) *demands* that the national government and its organisations establish laws for extended responsibility of producers of packaging materials, strict waste separation and waste reduction and finally 4) defined the aim of municipal waste reduction. It becomes clear that even before the reactor accident of 2011, the municipal policies of Town A were committed to climate protection. Interestingly, however, this was not translated into a focus on renewable energy but into strict municipal waste management. One reason for this is that before the implementation of the national feed-in tariff system renewable energy generation in Japan was too risky and required too high an investment. Before 3.11 the government was clearly prioritising nuclear energy, and it was very difficult to finance renewables.

More than the global outlook of the declaration suggests, however, the starting point of climate activism in Town A was not so much a concern for global environmental risks, but very local conditions and problems. The town is situated on a plain between mountain ridges, which is characterised by rice paddies and small canals. In an interview, the head of the town's environment office explained that the local farmers used to burn the remaining rice husks and other organic waste at the end of every harvest. As rice can be harvested several times a year, smoke clouds were hanging over the plain from summer to autumn. Furthermore, citizens disposed of their waste by throwing it in the irrigation canals between the rice paddies, which polluted the water and inhibited drainage. Being fed up with this, the former mayor ordered the city to develop a comprehensive waste reduction and management plan.

By recycling rice husks, sewage and other organic waste into biogas and fertilizer, the town was able to create new jobs, reduce pollution and waste treatment costs and supply local farmers with inexpensive fertilizer, fuel and electricity. Japan's generous FIT-scheme of 2011 thus did not play an essential role in the climate mitigation measures of Town A as important investments and strategic system decisions had been made before.

District b – demographic ageing, economic stagnation and political dependence

While Town A resisted repeated municipal mergers to maintain its autonomy and creatively used this freedom for gradually developing a far-reaching sustainability strategy, District b lost its autonomy in 2005. While A was able to stabilise its population also by using sustainability policies for creating new jobs and a positive dynamic, District b is continuing to age and depopulate. Since the merger, no major public plans and projects can be carried out here without the consent and tight control of the municipal government of City B. Paradoxically, it is precisely this remoteness and the lack of development pressure that has preserved many of the rich historic assets.

Through an initiative of local residents of District b and Kyoto University researchers in 2012, a 10.7-hectare area has been designated as Preservation District for Groups of Historic Buildings. Although this project is implicitly also highlighting the region-specific wood constructions and promoting the continuity of local craftsmanship as well as the use of local forestry resources, it reflects the growing consciousness in Japan in recent years regarding the conservation of historical buildings and landscapes. The area is part of a more conventional tourism strategy of City B, seeking to commodify historical sericulture farmhouses, temples, other pre-modern structures, and the newly designated Geo Park for attracting visitors. Additionally, in the historic core area, a thermal bath has been built that serves local residents as important community centre and a shop that sells and promotes local agricultural and craft products. Also a charging station for electric vehicles has been installed at the square, but this serves more like a display, with limited use and is not part of a comprehensive strategy to electrify mobility.

Besides official revitalisation plans and in the absence of *concrete* and far-reaching sustainability strategies in City B, several smaller, but disjointed citizen initiatives have sprung up that are mostly driven by young local people, or those who have moved here from urban areas. This is a familiar trend that we see happening all over Japan (Cf Dimmer 2017). A local young man in his late twenties, who returned to his hometown after having worked in Nagoya for a non-profit organisation dealing with regional revitalisation, developed one such project. He is also studying environmental studies at the prefectural capital, and scholars of his university support sustainability and revitalisation projects in the community. The organisation has 40 members and set up the *City B District b Firewood Cooperative* that seeks to utilise local forest resources for CO_2 neutral heating and energy generation. The objective of the cooperative is to contribute to saving energy and increasing self-sufficiency, the creation of jobs, thinning and up-keeping of the fragile mountain forests, and fostering biodiversity and CO_2 absorption. The group also wants to revive traditional charcoal-making skills that were once cultivated in the region. Like in many places in Japan, the local forest industry began to decline since the late 1960, when wood was increasingly imported and the Japanese forests were no longer carefully tended to.

Like many non-profits in Japan, the firewood cooperative suffers a lack of resources and professionalism. It operates, for example, only two small wood slicers and has no solid structures for storing the wood logs. Like most community organisations, it solely relies on volunteer work power in a country that is famous for its long working hours. Currently, the initiative is also caught in the dilemma that there is very little demand for firewood in the region as few house owners have wood boilers or wood stoves and thus there is no market for the firewood.

In order to scale up operations and make operations economically sustainable, a greater demand for this kind of forest resources needs to be created. One option discussed is to use wood for heating public buildings, but in order to retrofit these, time-consuming negotiations with officials in City B are needed, from which the organisers shy away. Furthermore, there is concern that firewood could harm the business of the only local filling station in the region, which now provides heating oil for the whole community.

Since 2015 City B has granted subsidies for the purchase of wood stoves in the hope to promote heating with locally produced charcoal, wood chips or pellets. However, as of 2016 most of the wood fuel comes from outside of the region since forestry has been declining over the last decades and there are no concrete strategies yet to revive it. One example of the piecemeal character of the current disparate sustainability strategies can be studied in the neighbouring city of B, where in 2015 an internationally known maker of construction vehicles has opened a 3,200 kW biomass plant that supplies electricity and heat for the production. The project, proudly advertised by prefectural officials as its showcase project, uses however mostly biomass from other regions of Japan and even abroad and thus contradicts any positive climate contribution through the enlarged environmental footprint. This too represents larger trends. Since 3.11 the government has offered lavish subsidies for promoting renewables. However, often the results are large-scale facilities, operated by big corporations that are not part of a wider sustainability strategy, that often create more local problems than they solve, and where little local value is created.

Only in July 2011, shortly after the nuclear accident in TEPCO's Fukushima Dai'ichi power plant in March that same year, City B launched a plan of local measures against global warming, which was based on a 2009 municipal declaration. The municipal government proclaimed the goal of reducing CO_2 emissions of base year 1990 by 130.800 tons annually until 2020. In March 2016 the city issued its 3rd Basic Environment Plan, but already in its 2nd Plan it announced the promotion of renewable energies, energy efficient buildings and buying energy efficient electric devices. Examples are subsidies for the installation of private woodstoves, which would contribute to "fighting global warming" and to a "more effective use" of forest thinning and wood waste. The municipality offers to subsidise 50 percent of installation costs, with up to 80,000 yen (~$700). Furthermore, the installation of PV on private houses is subsidised with 10,000 yen (~$90) for devices with a power between 1 to 2 kW and up to 50,000 yen for devices over between 5 and 10 kW. Small wind power is subsidised with 60,000 yen and solar heating with 30,000 yen per instalment. Aside from these subsidy schemes, a more concrete sustainability strategy, clear quantitative targets and a monitoring system are lacking.

In District b, however, these plans and measures have so far not caught on. While the area is abundant with natural energy resources such as wood, wind, rivers, sunlight and geothermal energy, social factors seem to hinder more progressive steps towards an energy transition. These factors are a lack of political autonomy, lack of knowledge, vision and leadership, as well as lack of economic capital and sunk costs due to the existing fossil fuel-based social and economic infrastructure.

Lacking political autonomy, the district is unable to strongly promote energy efficiency and reduce energy use in its public facilities such as its school building, without the discretion of the municipal government of City B, which only allows for investments in public infrastructure according to a fixed schedule. The school building is not scheduled for renovation for the next 10 years.

Furthermore, a lack of knowledge around renewables is contributing to a negative view of solar power and wind energy. The area is prone to heavy snow falls from December to March. Thus it was argued solar panels cannot be profitable as they would be covered in snow for too long. However, experiments haven't been undertaken yet. A proposal to use panels with a slippery surface that prevent snow from piling up on rooftops was met with the concern that this could endanger pedestrians in the narrow streets. Despite an immense geographic potential for wind turbines, wind was not exploited because according to a local leader "a scientist" had told the municipal government that wind energy was not reliable. This statement was then enough to discourage locals from further looking into the issue. As wind turbines require a major investment, it is maybe not surprising that a local community of 900 people has difficulties imagining these here. However, the commercial use of wood, which requires lesser investments, creates jobs and is naturally abundant, was also met with concern.

The residents of District b rely heavily on fossil fuels for cooking, heating and especially melting the snow on their roofs in winter. Therefore, the roofs are equipped with water pipes that are heated with oil or gas stoves. This technology wastes a huge amount of heat and results in high local expenditures on imported fuels. Heating houses with stoves fuelled with locally forested wood, however, might bankrupt the local filling station, the district representatives argued, which meant that the locals would have to drive long ways in order to fuel their cars. While the municipal government of City B displayed an unenthusiastic and technocratic approach to climate protection, the citizen representatives of District b demonstrated significant concern about the need to address energy use and climate change, but lacked knowledge and initiative to develop an own strategy.

Besides the latent political conflict between the municipal government and the mostly elderly representatives of District b, a generational conflict and a strong gender imbalance could be observed among the residents of District b. While the younger men were taking more interest in climate protection, learning, technologies and sustainable regional development, the older generation seemed more reluctant. Women, on the other hand, were completely absent from the public sphere of deliberations and meetings during our field visit. While young women were not seen, elderly women merely served the male representatives with food and drinks. When addressing this issue, a young man from District b explained this gender bias with a sardonic remark: "The women work in the background, while the men get drunk in the foreground."

The gender division in District b was considerably more pronounced than in Town A, where women, though underrepresented in political offices, played a role in advancing the public agenda of sustainability, operating a recycling store and a rest stop as well as organising public seminars.

Different from the younger generation, the senior citizens of District b seem to have subscribed to a narrative of loss of autonomy and victimisation by external forces, such as national top-down policies and globalisation that obliterated the traditional economic base of sericulture, forestry and charcoal productions. In fact, the villagers' experience with development plans and national policies has so far not been positive. In the 1960s the whole village had to be moved up the river to make way for a hydropower dam that flooded the whole settlement. National policies and cheap foreign imports had furthermore destroyed the local forestry and silk production. Thus there seemed to be little trust in plans that are made outside the district's confinements and a sense of resignation. Against these attitudes of the mostly elderly community leaders, the younger people have to struggle in order to advance their progressive ideas.

Conclusions

Like many industrial nations, Japan has pledged to combat climate change and to significantly reduce the country's greenhouse gas emissions. Within the government's climate action framework, encouraging renewable energy initiatives by local governments and the private sector plays a key role. However, what is boldly stated on higher-tier government policy papers runs into deeper structural problems at the local level, where, as shown, municipalities often lack vision, initiative, knowledge or simply the capacities to implement these national objectives. Foljanty-Jost and Haufe remark that there has been a remarkable "participation boom" in Japan in the last 15 years, however that "the function of the widely-touted new modes of participation is primarily limited to communication" and does not really lead to greater empowerment (Foljanty-Jost and Haufe 2011, p. 89). Also DeWit recognises this capacity gap of most local residents in Japan. Given the time pressure to act against climate change, he argues for a more technocratic approach to local climate action, bypassing the citizens (DeWit 2014). Particularly the case of District b made this clear. Although the government has recently declared an "age of regional (re)creation (*chihô sōsei*)", in which local governments are encouraged to develop resilience strategies against demographic decline as well as economic and environmental change, case study 2 exemplified some of the capacity bottlenecks.

On the other hand, Town A demonstrated that despite these problems, active citizens and progressive local governments in Japan have long started to tackle the environmental and climate change challenges and have successfully invested in renewable energies, energy conservation, recycling, new mobility solutions or utilisation of local resources. Pioneering local governments like those of Town A have stressed the importance of climate action and renewable energies long before national government. Determined leaders developed their comprehensive local visions from the bottom up and were only later recognised by the central government. Takao notes that examples like Town A are rather the exception than the rule (2012). He points out that only the largest cities as well as prefectures have governing capacities to develop a comprehensive approach to climate adaptation, while most medium-size localities act in a piecemeal fashion, as we have seen in City B and District b. They act according to clear national-level guidance on climate change and merely translate these goals into local plans without developing a clear strategy for implementation (ibid. 767).

Reaching beyond the local scale, various crosscutting grassroots networks have formed to overcome some of the above-mentioned problems. Examples are the network Mayors for a Nuclear Power Free Japan that connects progressive municipalities across Japan, or Transition Japan, which is the Japanese chapter of the global Transition Town movement and encourages alternative, citizen-based sustainability projects outside of the official policy system. Furthermore, independent think tanks such as Institute for Sustainable Energy Policies (ISEP), or Japan for Sustainability (JFS) have formed to carry out research, educate, and disseminate sustainability innovations in and outside of Japan between community innovators, or like-minded progressive local governments that contribute to emanate such knowledge. However, in our two case studies, such supra- and inter-local networks did not play a significant role.

Our brief analysis showed that in order to accelerate the energy transition, it should be framed from a broader perspective that emphasises the co-benefits from local action. As seen in Town A, such initiatives not only contribute to climate protection but also have the potential to improve local fiscal conditions, stimulate the local economy, strengthen social cohesion, reinforce civic pride and increase the adaptive capacity and community resilience. These are arguments that are less abstract than a global phenomenon such as climate change, and they have the potential to convince less-active local governments like City B to take more concrete steps.

References

Amai, A. (2010). Chichôson gappei ni tomonau jichitai seiji dôkô nitsuite [Analysis of the local election about the municipal amalgamation]. *Monthly Review of Local Government*, 375, 1–45.

Andrews, W. (2016). *Dissenting Japan: A History of Japanese Radicalism and Counterculture, from 1945 to Fukushima*. London: Hurst et Company.

Avenell, S.A. (2010). *Making Japanese Citizens: Civil Society and the Mythology of the Shimin in Postwar Japan*. Berkeley and London: University of California Press.

Dawson, C. (2011). In Japan, provocative case for staying nuclear. *The Wall Street Journal*, October 28. Retrieved from www.wsj.com/news/articles/SB10001424052970203658804576638392537430156 (2017.09.01).

DeWit, A. (2014). Japan's radical energy technocrats: Structural reform through smart communities, the feed-in tariff and Japanese-style 'Stadtwerke'. *The Asia-Pacific Journal*, December 1, 12(48), No. 2, 1–14.

DeWit, A. (2015). Japan's bid to become a world leader in renewable energy. *The Asia-Pacific Journal*, October 5, 13(39), No. 2, 1–26.

DeWit, A., and Tani, T. (2008). The local dimension of energy and environmental policy in Japan. *Japanstudien*, 20, 281–305.

Dimmer, C. (2016). Place-making before and after 3.11: Emergence of social design in post-disaster, post-growth Japan. *Review of Japanese Culture and Society* (Josai University), Special Issue 'Design and Society in Japan,' XXVII.

Dimmer, C. (2017). Japan after March 11th 2011: Between swift reconstruction and sustainable restructuring. In: Yan, W. and Galloway, W. (eds.) *Rethinking Resilience: Adaptation and Transformation in a Time of Change*. Springer International, forthcoming.

Feldhoff, T. (2008). Landes- und Regionalentwicklung zwischen Wachstum und Schrumpfung: Regionale Disparitäten und räumliche Planung in Japan. *Japanstudien*, 20, 35–67.

Flüchter, W. (1998). Geografische fragestellungen, strukturen, probleme [Geographic questions, structures, problems]. In: Pohl, M. and Mayer, Ha.J. (ed.) *Länderbericht Japan* [Country report Japan]. Bonn: Bundeszentrale für Politische Bildung.

Foljanty-Jost, G., and Haufe, C. (2011). 'Partizipationsboom' in Japan: Zur (Un)Möglichkeit einer Bürgerselbstverwaltung [Participation boom in Japan: On the (im)possibility of a civic self government]. In: Chiavacci, D. and Wieczorek, I. (eds.) *Japan 2011: Politik, Wirtschaft und Gesellschaft* [Japan 2011: Politics, Economy, and Society]. Berlin: VSJF, 89–119.

Government of Japan (GOJ) (2015). *Submission of Japan's Intended Nationally Determined Contribution (INDC)*. Retrieved from http://www4.unfccc.int/submissions/INDC/Published%20Documents/Japan/1/20150717_Japan's%20INDC.pdf (2016.06.07).

Gov-online (2016). *Ima koso chihô sôsei* [It is time for regional creation]. Retrieved from www.gov-online.go.jp/cam/chihou_sousei/ (2016.10.11).

Huenteler, J. Schmidt, T. S., and Kanie, N. (2012). Japan's post-Fukushima challenge – Implications from the German experience on renewable energy policy. *Energy Policy*, 45, 6–11.

Kantei (2016a). *Naikakufu chihô sôsei suishin jimukyoku*, official website. Retrieved from www.kantei.go.jp/jp/singi/tiiki/index.html (2016.10.11).

Kantei (2016b). *Chihô sôsei kanren kôfukin* [Subsidies for regional creation]. Retrieved from www.kantei.go.jp/jp/singi/sousei/about/kouhukin/index.html (2016.10.11).

Lindner, R. (2014). Setsuden: Die Energiekrise und gesellschaftliche Stromsparanstrengungen nach Fukushima [Setsuden: the energy crisis and social electricity saving]. In: Chiavacci, D. and Wieczorec, I. (eds.) *Japan 2014: Politik, Wirtschaft und Gesellschaft* [Japan 2014: Politics, Economy and Society]. München: iudicium Verlag, 201–216.

MacDougall, T.E. (1980). Political opposition and big city elections in Japan, 1947–1975. In: Steiner, K., Krauss, E., and Flanagan, S. (eds.) *Political Opposition and Local Politics in Japan*. Princeton, NJ: Princeton University Press, 55–94.

METI, Ministry of Economy, Trade and Industry (2014). *Strategic Energy Plan (Provisional Translation)*. Tokyo Ministry of Economy, Trade and Industry, April.

METI, Ministry of Economy, Trade and Industry (2015). *Settlement of FY2015 Purchase Prices and FY2015 Surcharge Rates Under the Feed-in Tariff Scheme for Renewable Energy*. Retrieved from www.meti.go.jp/english/press/2015/0319_01.html (2017.01.11).

MIC, Ministry of Internal Affairs and Communication (2014). *Kaso chiiki jiritsu sokushin tokubetsu shochi hô to kako no kaso 3 hô no gaiyô [Act on Special Measures Concerning the Promotion of the Independence of Over-Emptying Regions and Overview of the Three Past Laws on Over-Emptying]*. Retrieved from www.soumu.go.jp/main_content/000288551.pdf (2016.10.04).

NKS, Nihon Keizai Shinbun (2016). *Chihô Sôsei' Yamamoto Kôzô – Hatsunyûkaku ["Creation of regions" Mr. Kôzô Yamamoto (New in cabinet)]*. Retrieved from www.nikkei.com/article/DGXLASFS02H91_S6A800C1M10800/ (2016.10.11).

Osiander, A. (2007). *Der Fall Minamata: Obrigkeit und Bürgerrechte in Japan nach 1945* [The Minamata case: Authority and civil rights in Japan after 1945]. Munich: Iudicium.

Reiher, C. (2008). Kommunale Gebietsreformen der Heisei-Zeit und lokale Identität: Das Beispiel der Kommune Arita-chō. *Japanstudien*, 20, 163–192.

Sorensen, A. (2002). *The Making of Urban Japan: Cities and Planning from Edo to the Twenty First Century*. 1st ed. London: Routledge.

Stijn, L. (2014). *The Clean Energy Sector in Japan – An Analysis on Investment and Industrial Cooperation Opportunities for EU SMEs*. Tokyo: EU-Japan Centre for Industrial Cooperation.

Takao, Y. (2012). Making climate change policy work at the local level: Capacity-building for decentralized policy making in Japan. *Pacific Affairs*, 85(4), 767–788.

Tsukimori, O., and Sheldrick, A. (2016). *As Japan's Oil, Gas, Power Use Stalls, Coal Imports Hit New Record*. Reuters UK. Retrieved from http://uk.reuters.com/article/uk-japan-energy-demand-idUKKCN0V30N6

Yamamoto, T. (1999). Emergence of Japan's civil society and its future challenges. In: Yamamoto, T. (ed.) *Deciding the Public Good: Governance and Civil Society in Japan*. Tokyo: Japan Centre for International Exchange, 97–124.

Yoshida, S. (1999). Rethinking the public interest in Japan: Civil society in the making. In: Yamamoto, T. (ed.) *Deciding the Public Good: Governance and Civil Society in Japan*. Tokyo: Japan Centre for International Exchange, 13–49.

6 Multi-level climate change planning

Scale, capacity and the ability for local action

Neha Sami

Acknowledgements

This chapter is based on work funded by Canada's International Development Research Centre (IDRC) and the UK's Department for International Development (DFID) through the Collaborative Adaptation Research Initiative in Africa and Asia (CARIAA). Parts of this chapter have been previously published as an IIHS commentary on 'Climate Change and the Sustainable Development Goals' for the IIHS Urban Policy Dialogues 2016.

I would like to acknowledge valuable research assistance by Ritwika Basu and Amogh Arakali. I would also like to thank Amir Bazaz and Sumetee Pahwa Gajjar for very useful suggestions and for acting as sounding boards for ideas. The usual disclaimers apply.

Introduction

Cities and city regions across the world are emerging as critical actors in the struggle to deal with the challenges of a changing climate (Revi et al. 2014; Stone Jr. 2012; Bulkeley 2010; The International Bank for Reconstruction and Development/The World Bank 2010). Although climate change continues to be imagined as a global problem requiring solutions at international and national scales, there has been a growing involvement of municipal governments and other actors at the local (city) scale in dealing with challenges posed by a changing climate (Bulkeley 2010). This has emerged partly as a response to increasing frustration with international negotiation processes as well as the growing realisation that the impacts of climate change are going to be increasingly localised, and possibly amplified in urban regions (Stone Jr. 2012). There is also considerable investment that is being directed towards building capacity within cities not only by their own national and regional governments but also by international donor agencies, networks and institutions to help deal with these challenges.

This chapter examines local climate change action in Indian cities. Here, I understand 'local' as being at the settlement scale, focusing explicitly on urban settlements and their ability to respond to climate-related challenges. This focus is especially important given the scale of India's urban transition – by 2050, more than half the country will be urban (United Nations, 2014). As India continues its urban transition, urban populations in India are increasingly vulnerable to a range of environmental and socio-economic risks that will likely be exacerbated by climate change (Stone Jr. 2012; Mukhopadhyay and Revi 2009; Revi 2008). Planning and policy decisions taken now will have lock-in implications. In this chapter, I argue that there is a need for localised climate action at the city-scale in India, which engages with specific concerns that cities and urban residents are facing and can be operationalised within the existing governance framework. India's growing urban footprint also has global implications for climate change adaptation and mitigation with the proportion of vulnerable populations in Indian cities increasing and as a growing Indian middle class aspire to more Western standards of consumption.

Focusing on environmental governance in India, and especially on two states, Karnataka and Tamil Nadu, and their capital cities, Bangalore and Chennai, this chapter will examine how, if at all, planning for climate change (either adaptation or mitigation, or both) is being localised, and what challenges and opportunities exist. The emphasis here is particularly on government efforts to plan for climate change challenges. While the aim of this chapter is to begin to identify how Indian cities can plan for climate change, I also engage in some detail with larger state-level planning processes. This is important because city governments and their officials in the Indian context rarely have the power to make their own plans (Weinstein et al. 2013) or make choices about particular development pathways. For the most part, decision-making power rests with state government departments, with city governments being largely implementing agencies. Therefore, to know how urban governments can address the climate change challenge, it is necessary to first understand how the state government departments are engaging with the issue.

This research is part of a larger project (Adaptation at Scale in Semi-Arid Regions or ASSAR) funded by Canada's International Development Research Centre (IDRC) and the UK's Department for International Development (DFID) through the Collaborative Adaptation Research Initiative in Africa and Asia (CARIAA).[1] Working in seven countries in semi-arid regions, this project seeks to understand the factors that have prevented climate change adaptation from being more widespread and successful, and the processes – particularly in governance – that can facilitate a shift from ad-hoc adaptation to large-scale adaptation. The study sites in India under the ASSAR project are based in the states of Maharashtra, Karnataka, and Tamil Nadu. This chapter is focused on two of these three states: Karnataka and Tamil Nadu.

Drawing on interviews with local government officials, academics, experts and other knowledgeable observers, academic writing as well as reviews of policy papers, plans, reports and news articles, this chapter explores questions of scale, engaging with regional and local level actors; capacity; and the ability to act and implement solutions.[2] I find that, in most cases, there is no specific institutional architecture, either within government or outside it, to deal with challenges of climate change beyond the level of the state governments.

As this chapter will show, the lowest scale at which climate change planning takes place in India is that of the state. Using the national plan as a blueprint,

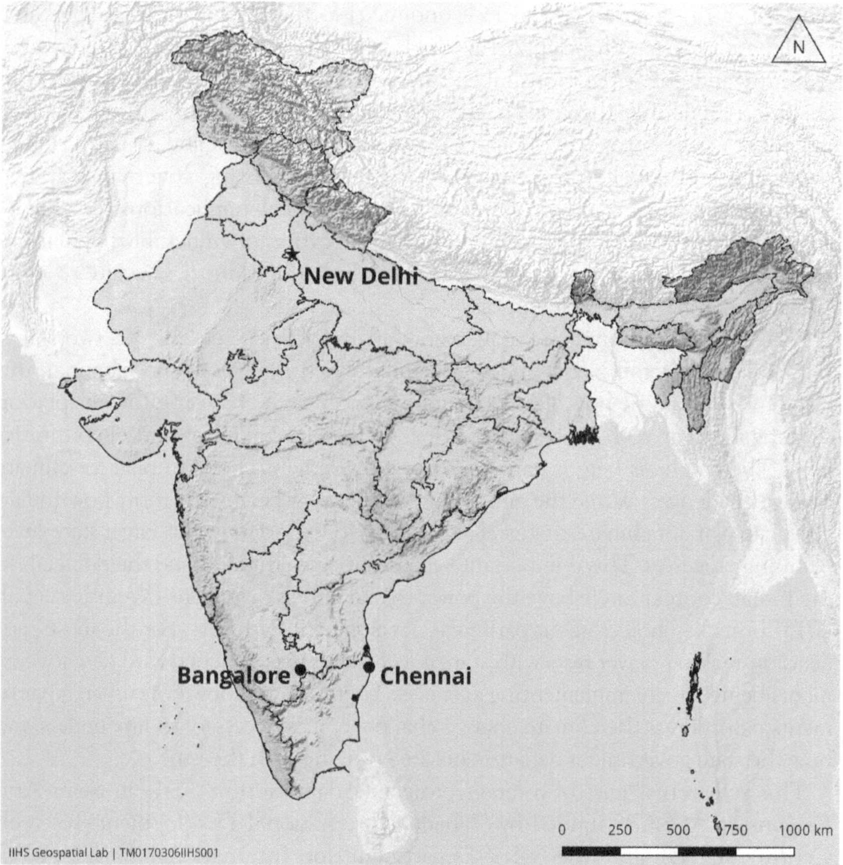

Figure 6.1 Case site locations

Source: IIHS Geospatial Lab 2017; Administrative Atlas of India 2011; Natural Earth 2017

state plans aim to integrate national priorities with the respective state's development goals. Government officials from urban local bodies are typically absent from this plan-making process. Moreover, since these are state plans, the focus is much wider. Plans tend to focus on an extensive range of issues from agricultural productivity and livestock management to regional-level initiatives such as a state-wide shift to cleaner energy, as in Karnataka. Plans mention urbanisation and cities to varying extents, depending on local and regional context. In Karnataka and Tamil Nadu, the state action plans mention urban issues, highlighting how climate change will exacerbate many existing problems and concerns. However, there doesn't seem to be much engagement in either of the two plans with local agencies, nor is there a clear set of actions at the city scale that would enable implementation or a clear plan for building capacity in urban governments to deal with climate change challenges.

The next section gives a brief overview of the environmental governance framework as well as the institutional structure in India. The rest of the chapter focuses on the two states and cities. The chapter concludes with a section on challenges and opportunities, especially at the local scale.

Environmental governance in India

In the Indian context, environmental governance has largely been underpinned by concerns around reconciling developmental priorities and economic growth with sustainable development pathways (Dubash and Joseph 2015; Nair 2015). Since the 1970s, India has maintained a relatively consistent stance internationally emphasising that developing countries need to focus on developmental priorities such as poverty alleviation while simultaneously addressing national and global environmental challenges (Dubash and Joseph 2015; Revi and Sami 2013; Revi 2008). There have been several national-level policies, legislations and programmes that the Indian government has put in place to deal with environmental concerns, including the creation of a series of 'end of pipe' regulatory agencies, pollution control legislation and the passage of several landmark court judgments. Various citizen-led initiatives have also been instrumental in the formation, implementation and enforcement of environmental policies. For example, in May 2016, the National Green Tribunal fined a real estate developer for encroaching on and attempting to develop in and around wetlands in Bangalore (Bangalore Mirror Bureau 2016). This decision emerged from a review of a Public Interest Litigation (PIL) that several Bangalore-based citizen groups had filed to demand protection of lakes and water bodies in the city.

The integration of environmental and development concerns can be traced back to the 1960s and '70s, when the Five Year Plans clearly spelt out the need for assessing and addressing environmental problems to enable national development goals. In 1985, an independent nodal national-level ministry was set up – the Ministry of Environment and Forests, now renamed the Ministry of Environment, Forests, and Climate Change (MoEFCC), has adopted an overarching environmental policy (the National Environmental Policy 2006), which sets mandates for concerned departments, both at the apex and at the sub-national (state and provincial) levels for implementation of environmental quality standards within their defined areas of jurisdiction.

Most states also mirror this structure – there are typically state departments of environment that carry out the mandate from the national government, although the specifics of the institutional structures, governance mechanisms, and functions differ from state to state. However, there is very little understanding of urban environmental issues within the state departments, and the approach to dealing with these remains largely reactive. In addition, there is no designated agency at the city level that has been given the responsibility of dealing with environmental issues. There are of course several aspects of environmental management that are distributed among various local agencies – for example, water and sanitation as well as solid waste management – but there is little coordination between the different agencies in charge.

Urban development and environmental issues (including climate change) are handled by two different ministries at the national level. There is very little coordination between both ministries and their departments whether at the state or city scale. Moreover, the institutional structure at the state level itself is usually not very robust. There is also no specific agency that has the mandate to deal with climate change issues at the local level. Awareness, and the capacity to deal with the challenges are also issues. There is little integration of environmental sustainability issues more broadly and climate change specifically in the urban planning process in most Indian cities. Non-state actors also have different levels of engagement with planning and decision-making processes, whether to do with environmental issues or urban development more broadly (Sami forthcoming; Sami 2013; Ghertner 2011; Kundu 2011).

In addition to these, there are two further challenges as far as localising climate action at the city-scale goes. The first is the lack of power at the city-scale: while the Indian national government has passed legislation that requires decentralisation, the implementation of this has been piecemeal and patchy – most city governments do not have control over urban planning processes or over municipal finance. To be able to proactively plan to tackle climate change at the urban scale, cities will need the ability to raise finances and decide how the money is spent. Cities and their governments also need to be able to draw on and incorporate expertise and knowledge from a range of non-state actors, including academics, civil society groups and the private sector. The second is the lack of capacity to act: city-level (and often even state-level) officials are often handicapped by the lack of capacity and technical knowledge to be able to understand the problem and begin formulating responses to it.

Climate change in Indian cities is rarely seen as a challenge that can and must be tackled at the urban scale by city and state governments (Sharma and Tomar 2010). This is exacerbated by the governance framework in India, where local governments have very little power and decision-making authority (Weinstein et al. 2013; Pinto 2000). Urban governance is also very fragmented, and responsibilities for urban planning are divided among a very wide range of agencies, often leading to a lack of ownership over larger issues that demand an integrated approach. In the case of the Bangalore metro region, for example, there are at least five different agencies managing different aspects of water resource management, from water supply and sanitation to managing and protecting the city's lakes, but coordination between the different agencies remains an issue.

There is also a huge challenge within government in terms of the awareness of climate change itself, its impacts in urban areas and the capacity to engage and deal with it. That said, there are examples and instances of what Pasquini et al. (2015) call 'environmental champions' in a few Indian cities (Surat and Shimla are two examples): political leaders who have explicitly engaged with climate change as an issue at the urban scale and attempted to develop strategies to tackle emerging concerns (Pasquini et al. 2015). These remain a minority, although there is a growing awareness and a slow rise in local officials willing to engage with the issue. For the most part, however, local governments and government officials do not acknowledge climate change as an issue that can be dealt with at the local scale.

However, there is a gradual change emerging in this perception, helped by significant domestic and international non-state actors, ranging from donor agencies and foundations to civil society groups that have been working closely with city as well as state governments to build awareness and capacity within government agencies. Several interview respondents involved in training government officials indicated that requests for training on specific adaptation-related issues had increased from state governments over time. Those who had been involved with state governments in an advisory capacity also indicated that while there had been early resistance to engaging with environmental issues, this has changed as international and domestic policy has evolved. In Surat, for example, the Surat Climate Change Trust has emerged from the second phase of the Asian Cities Climate Change Resilience Network (ACCCRN) funded by the Rockefeller Foundation. The trust, which is one of the first city-level multi-stakeholder trusts in India, includes members from state agencies, the private sector and various academic institutions. It has taken the lead to implement several ACCCRN projects as well as a range of interventions outlined in the City Resilience Strategy (Surat Climate Change Trust 2015).[3] However, interview respondents emphasised that while these efforts are important, they need to be backed by strong political will and consistency in order to be effectively implemented.

Climate change policy and decentralisation

Climate change policies in India are firmly anchored within a co-benefits framework, focusing on leveraging the synergies between development and climate outcomes (Dubash 2012; Revi 2008). These were first articulated in the National Action Plan on Climate Change (NAPCC) 2008 and were also explicitly stated in the 12th Five Year Plan, introducing measures that promote economic development while yielding secondary climate benefits. Prior to 2007, the domestic institutional structure was skeletal at best – the overriding concern was a developmental one, with climate policy being "synonymous with foreign policy on climate change" (Dubash and Joseph 2015, p. 11), which was reflected in domestic policy approaches. Since 2007, however, there was a gradual shift internationally around climate change debates, starting with the Bali Conference of Parties (COP) in December 2007 and the Kyoto protocol, leading to the COP in Copenhagen in 2009, and most recently the 2015 Paris Agreement. As pressure for climate action grew globally, it also led to growing domestic attention on the issue, and the creation of a domestic climate change policy.

The NAPCC, released in 2008, aimed to create a directional shift in India's development trajectory by integrating climate concerns with larger developmental ones (Dubash and Joseph 2015; Byravan and Rajan 2012) through sustainable development pathways that advance both economic and environmental objectives. The plan itself was organised around eight Missions or sectors, which map out long-term, integrated strategies towards national goals around climate change, ranging from solar power development to agriculture and water as well as a separate Mission on the Himalayan ecosystem (Byravan and Rajan 2012; Prime Minister's Council on Climate Change 2010).[4]

Following the development of the NAPCC, the Indian national government asked state governments to mirror this process at the regional level, focusing on developing policies and plans that align with the eight NAPCC Missions as well as the development priorities of the specific state. The aim was to decentralise the implementation of the eight NAPCC Missions beyond the national scale (Dubash and Jogesh 2014). As of October 2016, 32 out of 37 states and union territories in India have State Action Plans on Climate Change that have been endorsed by the National Steering Committee on Climate Change (Ministry of Environment and Forests 2016). A review of five State Action plans (including Karnataka but not Tamil Nadu) by Dubash and Jogesh (2014, p. 87) finds that while the process of developing these plans has opened up the conversation around issues of climate change and environmental sustainability, there has not yet been any opening for transformational change or the 'directional shift' in India's development pathway that the NAPCC was aiming to create.

Urban climate change and urban developmental challenges are reflected to varying degrees in the different SAPCCs. The state plans for Karnataka and Tamil Nadu, for example, do discuss urban issues and challenges, given the importance of urban regions to both states. However, they do not go much further beyond identifying issues and proposing broad interventions. Similarly, the SAPCCs address issues of vulnerability or risk mitigation in urban areas to different degrees. It is rare though for urban local bodies – like the city corporation or the planning agency of the city – to be involved in the formulation of these plans, or be part of the implementation process.

The mandate for urban planning and governance in India is fractured – the responsibility for these functions is typically distributed across a range of city, parastatal and state-level agencies. Municipal corporations in Indian cities are often financially constrained and understaffed and lack the technical capacity to undertake planning and development, responsibility for which often lies largely with state government-appointed parastatal bodies like the development authorities.[5] Moreover, there is little clarity at the state and local level on the larger environmental governance agenda and urban local bodies are unsure of where the responsibility for environmental issues lies and the extent to which they are responsible.

It is within this larger context that I examine the climate change action plans and the process by which they were put together in Karnataka and Tamil Nadu.

Planning for climate change: Karnataka and Tamil Nadu

About half of Tamil Nadu and 40 percent of Karnataka is urban (Census of India 2011), although both states have very different patterns of urbanisation.

Karnataka's urban population is dominated by Bangalore, its capital city. Tamil Nadu, on the other hand, has a more dispersed urban population with several smaller urban centres as well as the large Chennai urban region. As the urban population in both states increases, its vulnerability to a range of environmental issues has also grown – Karnataka is struggling with drought and water scarcity for industry and agriculture as well as rising temperatures and urban heat islands in its cities. While Tamil Nadu also faces similar issues, its cities are additionally at threat from sea level rise and flooding. There is also a growing proportion of their urban population that is marginalised and increasingly vulnerable to social, economic, and environmental risks, which climate change will exacerbate.

Like most other Indian cities, Bangalore and Chennai do not have individual climate action plans but rather form part of the overall state strategy for dealing with climate change. This section focuses on how each of these states is dealing with climate change. I review the climate change action plans that both states have developed and the National Steering Committee on Climate Change has endorsed, the process by which these were developed, and how, if at all, they are being implemented.

Planning for climate change: Karnataka

Located in south-western India, Karnataka is among the most urbanised states in the country, with about half of its GDP coming from the service sector, located largely in its capital, Bangalore. The economic profile of Karnataka, and Bangalore in particular, has changed over the last two decades, transitioning from manufacturing (especially textiles) to service sectors, largely information technology and related services. Possibly the most critical environmental issue in the state is that of water security in both urban and rural areas. In fact, a major flashpoint between Karnataka and Tamil Nadu is the sharing of water resources from the Cauvery River that flows through both states and has been the centre of many violent confrontations in both states – the most recent of which was in September to October 2016 (Janakarajan 2016). This is compounded by the dramatic decline in lakes and water bodies over the last two decades in Karnataka, especially in Bangalore, as a result of rapid economic and urban growth.

Although the state possibly has "the strongest independent scientific and research capacity on climate change" among other Indian states (Dubash and Jogesh 2014, p. 1), and there is an acknowledgement within higher levels of the state government of the seriousness of climate change, it has yet to feature prominently in Karnataka's development agenda. The official State Action Plan for Climate Change (SAPCC) draws on three collaborations that resulted in three separate plan documents for the state. Only one was officially recognised and approved. As part of the research for this project, we reached out to members from all three groups with interview requests. However, we were only successful in meeting with the Environment Management and Policy Research Institute (EMPRI) – the official nodal agency, and officials at the local planning agency – the Bangalore Development Authority (BDA).

The three plans

As part of the SAPCC process, three different plans were prepared for Karnataka. Two of these were undertaken by independent research agencies or think tanks, and only one (prepared by EMPRI) was the officially endorsed plan. The plans were prepared sequentially with the Bangalore Climate Change Initiative's plan being the first, followed by the official EMPRI plan, and the third was made by the Centre for Sustainable Development (CSD).

The plan prepared by the Bangalore Climate Change Initiative–Karnataka (BCCI-K), was one of the earliest efforts in the climate planning process. The BCCI-K was a coalition of the various leading scientific and research institutes in the state. Their plan was focused more on the scientific assessment of the possible climate change impacts for Karnataka because the coalition did not believe that EMPRI had the capacity to do this (Jogesh and Dubash 2014). The BCCI-K plan also had international technical support from Lord Stern and the India Observatory at the London School of Economics (LSE) and financial support from the World Bank (Jogesh and Dubash 2014, and personal interviews). However, their interaction with government agencies and other stakeholders beyond the academic community seems to have been limited. The plan that BCCI-K proposed therefore functioned largely as a resource for the plans that followed. The subsequent two plans, including the official plan, drew heavily on the BCCI-K's scientific assessments (ibid.).

The official Karnataka State Action Plan on Climate Change (SAPCC) was prepared by EMPRI, the nodal agency at the state level to prepare and subsequently to facilitate implementation of identified projects. EMPRI comes under Karnataka's department of Forest, Ecology, and Environment, and was appointed by the State Committee on Climate Change to prepare the SAPCC for Karnataka (Jogesh and Dubash 2014). This was different from other states in the country. Rather than appoint an external consultant (like GIZ for Tamil Nadu), the state government of Karnataka preferred to appoint a government agency like EMPRI to prepare the plan. This was mainly because it was felt that EMPRI would find it easier to convene meetings and gather information across different government agencies than a private consultant (Jogesh and Dubash 2014; personal interviews). This plan drew heavily on the earlier BCCI-K plan for scientific and technical data as well as several other secondary sources. However, as they told us in interviews, EMPRI officials also conducted a series of meetings with various stakeholders in different government agencies, including some representatives from local city agencies in Bangalore like the Bangalore Development Authority (BDA – the city's chief planning agency), and the Bangalore Electricity Supply Company (BESCOM).[6]

In addition, EMPRI conducted structured interviews with various government departments to understand perceptions around climate change, threats from climate change and activities undertaken to mitigate these threats (Jogesh and Dubash 2014). Officials at EMPRI explained that conducting stakeholder meetings had been challenging, particularly because there was a lack of understanding about climate change and its impacts, especially in urban local bodies. EMPRI then drew on The Energy Research Institute (TERI) for assistance in synthesising findings from different sources and compiling it as one document. The Karnataka plan covers a wide range of sectors from agriculture and water to health and biodiversity and also identifies synergies with eight NAPCC Missions across each of these sectors.

There are references to urban development and growth through the entire plan, identifying these as future challenges (both in terms of risks and opportunities) to sustainability in the state, especially from a rural development perspective (in the case of agriculture and water resources, for example). There is also a chapter dedicated to Urbanisation in the Karnataka SAPCC, which identifies four target sectors as part of the action plan: Urban Planning and Land Use, Transport, Waste Management and Housing. It lays out the current state of each sector and identifies various on-going activities in each of these sectors that are part of state-level initiatives (rainwater harvesting and solar water heaters in buildings, switching to greener fuels for public buses and improved waste management practices, for example). These are aimed at urban regions more broadly, not aimed particularly at Bangalore.

The Karnataka SAPCC also identifies activities under the NAPCC that may be relevant for these sectors, but these are unfortunately much too broad and lack specific targets or actions associated with them. The focus of this plan was mainly on aligning Karnataka's state plan with the overall NAPCC goals, but there was no attempt to take this forward, or build linkages between various activities or plans at the state level and the goals of the NAPCC Missions (Jogesh and Dubash 2014). There was also little done to foster broader engagement around the issues that the state was facing. The EMPRI SAPCC process aimed to be exhaustive in reaching out to government departments and agencies, but made little attempt to engage non-governmental stakeholders, including activists, academic institutions (apart from their use of the BCCI-K report) or the private sector.

The third plan was prepared by an independent think tank based in Bangalore called the Centre for Sustainable Development (CSD). As of October 2016, the plan has not been made public, and repeated requests for meetings or the plan document have received no response. The reason why a third plan was deemed necessary or what CSD's mandate was in the preparation of the plan seems unclear (Jogesh and Dubash 2014).

Implementation

The Karnataka SAPCC identified a broad set of activities by sector in order to implement the plan's recommendations. However, almost none of these have actually been operationalised almost three years after the plan was submitted for approval. There are five pilot projects that EMPRI is currently running in test sites around the state with a small amount of funding from the government of India. These focus on agriculture, livestock management, health impacts of climate change, biodiversity and forestry, none of which are in urban areas. EMPRI officials explained that they planned to test solutions in smaller sites before expanding coverage to other parts of the state.

When asked about urban areas, officials indicated that some efforts were being made in Bangalore, such as mandatory rainwater harvesting, incentivising solar water heaters, a plastic bag ban and the opening of the first phase of the Bangalore Metro. They also emphasised that there may be other activities that could have co-benefits but weren't being identified as such. None of these is targeted particularly towards climate change, but they are good sustainable development practices. EMPRI also had very little engagement with any local Bangalore agencies, which seems to point to a lack of coordination between the state and local levels. During our interviews, there also did not seem to be a sense of urgency around the issue, but rather that activities needed to be thought through carefully before implementing. It was also incumbent upon individual agencies to incorporate climate action in their agendas. EMPRI's main focus at this time was on conducting training workshops on a range of environmental issues for government officials in Karnataka.

EMPRI officials pointed out that one of the biggest bottlenecks they faced in implementation was the lack of funding. While the national government had mandated the creation of the SAPCCs and aimed to decentralise the implementation of the NAPCC missions, there was little by way of financing that was made available to state governments or their agencies. They have been encouraged to look for other sources of funding that can enable the implementation of activities recommended in the SAPCCs. Potential sources include international agencies and donor organisations as well as private sector institutions. This has made it even more challenging for agencies like EMPRI, which is a research organisation and lacks the autonomy to raise and allocate funds. Without dedicated financing, it is also difficult to incentivise local government agencies such as the Bangalore Development Authority, the city's chief planning body that are not as invested in these questions as agencies like EMPRI.

This is compounded by the fact that government officials across multiple departments lack the technical capacity to tackle environmental issues in general, and climate change in particular. EMPRI's current activities focus more on developing training programmes for state government officials from a range of departments and ministries. These programmes are targeted largely at creating awareness and building capacity within state government. Participation in these is often mandatory and not always demand-based. As a result, the impact and efficacy of these programmes is somewhat lessened since participants are not always invested in the issue and are not often clear on how to take this forward.

Planning for climate change: Tamil Nadu

Tamil Nadu is one of the most urbanised states in the country, with almost half its population living in urban areas. Since the mid-1990s, Tamil Nadu has been one of the fastest-growing states in India as well as one of the most industrialised (Vijayabaskar 2010). Urban regions are very important to Tamil Nadu's economic and spatial growth. Several of these are located along the coast and are increasingly vulnerable to a range of climate-induced risks, such as sea-level rise, increased cyclonic activity and flooding. These risks are exacerbated by poor urban planning and unsustainable land use management practices. It is therefore not surprising that urbanisation features as both a risk and an opportunity in Tamil Nadu's State Action Plan on Climate Change (SAPCC).

The Tamil Nadu planning process followed a very different route than Karnataka's. Unlike in Karnataka, only one plan was prepared. Tamil Nadu also drew on external consultants for planning support and expertise: GIZ, the German Development Agency, worked with various government stakeholders to produce the SAPCC. In addition, it drew on and aimed to integrate with Vision 2023, the strategic infrastructure development plan that had been produced in 2012 (Government of Tamil Nadu 2014).[7] The SAPCC therefore aligns itself with these aims, while simultaneously laying out a plan to tackle climate change. The plan also explicitly states its intention to reduce vulnerability and exposure to risk, focusing largely on adaptation strategies and leveraging mitigation co-benefits wherever possible. These will not be implemented in isolation but "integrated into the regular developmental planning process, keeping in tune with the convergence principles articulated in the State's Twelfth Five Year Plan" (Government of Tamil Nadu 2014, p. 15).

The State Climate Change Steering Committee appointed GIZ to assist with the plan preparation process. Following an inception workshop in 2010, seven focal areas that mapped onto the NAPCC missions were identified.[8] Working groups were created for each of these, which were tasked with preparing sectoral papers with the assistance of GIZ. These were reviewed and integrated to form the SAPCC for Tamil Nadu. The plan lays out the key risks that the state faces, mainly from rising temperatures and increased cyclonic activity, and recognises a set of challenges within the seven sectors identified earlier. For each of these sectors, a set of activities has been proposed to help tackle the challenges identified and mitigate some of the risks together with budgets needed for implementing these activities for each sector. It also lays out a clear implementation and governance mechanism: the nodal agency for implementing the SAPCC will be the Tamil Nadu State Climate Change Cell (TNSCCC) under the Department of Environment within the Government of Tamil Nadu. A Steering Committee will also be constituted with representatives from all relevant line departments within the state, headed by the chief secretary. All sectoral line departments within the state as well as any other relevant agencies were also tasked with setting up climate cells that would then coordinate with the State Climate Change Cell to operationalise and implement the SAPCC.

In terms of urban issues, the SAPCC has a section on Sustainable Habitat that mirrors the NAPCC mission. There are also references to challenges posed by rapid urbanisation throughout the plan, particularly as almost half of the state's population lives in urban areas, and this is only likely to increase. Within the section on Sustainable Habitat, nine sub-sectors have been identified, including housing, drinking water, health and sanitation, urban development, waste management, transport, energy, pollution and greening of urban spaces. In each of these sectors, particular challenges and existing as well as proposed activities have been identified together with specific line departments within the state government that deal directly with these issues. The plan identifies both adaptation and mitigation measures as part of the implementation strategy. As with Karnataka though, urban local bodies are largely absent from both the planning and implementation processes as laid out in the plan. This is somewhat surprising in the case of Tamil Nadu because Chennai is one of the few cities in India that had (until recently) a directly elected mayor and the state government had actually made an attempt to decentralise urban governance across cities and towns in the state in keeping with the 74th Constitutional Amendment Act (National Institute of Urban Affairs (N.I.U.A.) 2005).

Despite creating an elaborate structure and governance mechanisms, as well as a list of activities outlined in the SAPCC, very little has actually been operationalised. The plan activities themselves seem to be a repackaging of already existing activities rather than using the plan development process to come up with innovative strategies to tackle the challenges and impacts of climate change. While the State Climate Change Cell seems to have been created, there is little information available on its activities, and its website is no longer functional. According to the Department of Environment's website, they are currently undertaking a climate change adaptation programme in collaboration with the German government and GIZ, but only in rural areas.[9] In the aftermath of the 2015 Chennai floods, there were renewed calls to focus on urban areas and develop strategies that would help urban regions in the state to deal with climate change issues, but it is unclear if the SAPCC has been revised to reflect this. Participation in the implementation process seems to be limited largely to state government agencies and their officials, although the plan does identify civil society, various private sector entities, as well as academic and research institutions as valuable resources. However, the plan has no clear strategy that indicates how they propose to engage these stakeholders.

Risks and opportunities

Although Karnataka and Tamil Nadu differed in their approaches to formulating and developing the plans, there are nonetheless several similarities between the two plans. The SAPCCs for Karnataka and Tamil Nadu do pay attention to some of the climate-induced challenges and concerns that urban regions are beginning to face. However, their focus remains somewhat limited to already existing activities, rather than using these plans as opportunities for innovation. In both plans, city governments and other urban local bodies are largely absent. Local agencies and officials rarely participated in the planning process, or in the identification and prioritisation of activities in the SAPCCs. Moreover, since the focus of the plans was at the state level, specific cities and their issues did not get reflected.

States are increasingly turning to consultants to help with planning processes. Karnataka decided to develop the plan in-house since it was assumed that other government agencies would be more willing to engage with a process that a state institution was leading, rather than an external consultant. Interview respondents repeatedly emphasised the importance of credible local long-term partners to help with developing plans as well as implementation. While there is no evidence from the State Climate Action planning process that plans made by external consultants (domestic or international) were more inclusive or effective, interview data indicates that there was more credibility if partners or consultants were local and embedded within the state or city context.

As cities across the world begin to demonstrate their leadership in climate action, Indian cities need to work out a mechanism to participate in this larger global movement. However, there are several challenges to enable this participation. The first, and perhaps the biggest, is the fragmentation of urban governance in India. This makes coordination across different agencies and scales difficult. Although state governments may devolve power over urban development and planning to local governments, most have chosen not to (Weinstein 2009; Pinto 2000). The responsibility for planning and governance functions is typically distributed across a range of city, parastatal and state-level agencies in both Chennai and Bangalore.

The power to control urban planning in Bangalore is dispersed widely among several municipal and state-run agencies, often leading to contentious decision-making as jurisdictions and functions of various agencies tend to overlap. While the state of Karnataka has a Disaster Management Authority with a policy for disaster management and a State Climate Action plan in place as well as district level disaster management authorities for the Bangalore metro region, they have not begun to operate effectively (IIHS et al. 2013). While there is agreement and buy-in at the higher levels of government, there is little understanding of these issues at the local institutional (operational) level (IIHS et al. 2013). For example, planners from the Bangalore Development Authority, the city's chief planning agency, indicated that this was not part of their mandate, nor did they perceive these as critical issues. Although the new draft master plan does incorporate an integrated planning approach and mentions the need to integrate with environmental plans being developed for Bangalore (Sami, forthcoming), this does not translate into capacity or understanding at the operational level.

Tamil Nadu, on the other hand, has been among the most successful states at decentralising urban governance (National Institute of Urban Affairs (N.I.U.A.) 2005).[10] Despite this, local agencies have been notably absent from the climate planning process in Chennai. Environmental management responsibilities have been moved to departments that are no longer within the city corporation, as a move to improve efficiency, reducing the corporation's capacity to deal with these issues (Tanner et al. 2009). In addition to the fragmentation of governance, there is a constant tussle between the two dominant political parties in the state for control of both the city and state government (Institute of Development Studies (IDS) 2007). As several interview respondents also explained, when the same political party is in control of both governments, there is little obstruction to planning and implementing development projects. However, when there are two different parties in the city and state government, the functioning of the city corporation slows down.

Climate and urban policy making and planning, including resilience planning, is concentrated within state governments. The role of urban local bodies, civil society and communities tends to be that of implementation. This leaves little room for innovation, and the potential for cities to foster inclusive and sustainable development will remain unrealised. India's current urban regulatory and governance structures are inadequate to address climate challenges. Although an appropriate urban-level platform is needed to engage with different stakeholders on various aspects of climate action, there are opportunities that exist within the existing structure that can and should be leveraged. One example of this is the formation of the Surat Climate Change Trust that brings together different stakeholders to help address environmental risk. This compact also emerged after a decade-long intervention through an internationally funded programme.[11] Surat also has a long history of environmental and public health risks that has made the local government as well as other city stakeholders aware of the importance of working collaboratively to address these issues.

From their experiences of working with Surat and other Indian cities, interview respondents indicated that it was important to identify a credible long-term local partner who could coordinate with government agencies and take leadership in knowledge management and implementation. In Gorakhpur, one of the other ACCCRN cities, there was very little engagement with the government, and a local non-profit led most of the ward-level resilience planning. The plans that were developed through this process, however, have been implemented only in a piecemeal manner, focusing on small projects that could be undertaken at the neighbourhood level. Interview respondents emphasised that the lack of engagement from government agencies is the biggest challenge in climate action in India. One respondent especially stated that it was preferable to have lower capacity but greater engagement, since that often enabled projects to get off the ground.

There are also several government of India schemes and related interventions that may be leveraged as part of a climate-action oriented agenda. For example, under the Atal Mission for Rejuvenation and Urban Transformation or AMRUT, close to 500 cities, including Bangalore and Chennai, will be funded with an investment of INR 50,000 crores targeted over five years. Some of the key thrust areas under AMRUT that every selected city has to engage with include the development of an integrated master plan, developing public transportation infrastructure focusing on non-motorised forms of mobility, sustainable water management and increasing green spaces within the city.

However, as this chapter has shown, three key issues need to be addressed to make localised city-based action possible. The first is the capacity to act: officials in local (city) agencies lack the knowledge and skills to tackle problems of this nature. Interview respondents within local agencies also said that it was one of the reasons governments were turning increasingly to consultants, since it was difficult to both build internal capacity as well as hire new staff. The second issue is the political authority to implement change at the city scale. As discussed earlier, although the 74th Constitutional Amendment decentralised power to local governments, its implementation has been piecemeal. As a result, few local governments have the power and authority to make decisions. Given this, as well as the fragmentation of powers across different local agencies, it becomes very difficult for city officials to implement policy or take decisions on spending, for example. Even in states like Tamil Nadu, where the 74th Amendment has been relatively better implemented, local agencies have been absent from environmental and climate governance processes. The third concern is the lack of willingness to act or engage. According to interview respondents, this was the biggest obstacle to successful policy implementation. There are few local officials who have engaged with environmental or climate-related issues – for the most part, city officials do not consider these to be part of their responsibility, and therefore choose to not engage. Civil society actors and international organisations like the Rockefeller Foundation who have worked with Indian city governments over the last decade or so have emphasised the need to build greater awareness among local city officials in order to increase engagement and consequently their willingness to act.

This cannot happen in isolation, however, or only at the city-scale: there needs to be a commitment from national and state governments to prioritise climate action (for both adaptation and mitigation) through financing, improved governance and coordination across scales and sectors and with the participation of multiple stakeholders.

Notes

1 For more information on the ASSAR project, please see: www.assar.uct.ac.za/, and for more on the CARIAA program, please see: www.idrc.ca/en/initiative/collaborative-adaptation-research-initiative-africa-and-asia

2 In this instance, I use 'capacity' to mean the knowledge and skills to be able to address environmental challenges. I use 'ability' to mean both the authority as well as the willingness to act.

3 For more on the Surat Climate Change Trust, see: http://scct-surat.in/Default.aspx

4 For a more detailed discussion of the NAPCC, please see (Byravan and Rajan, 2012). For an in-depth discussion of the institutionalisation of climate policy, especially at the national level, see (Dubash and Joseph, 2015)

5 'Municipal Corporations' in Indian cities are elected local governments that administer urban areas with populations over 1 million. Municipal corporations are responsible for urban planning, regulation of land use and buildings, public health and solid waste management, water supply, slum improvement and upgradation, among other issues. The specific services provided differ from city to city. However, there are typically a range of other urban local bodies (like the Water Supply Board, or the city Development Authority) that also undertake a range of functions that overlap with those of Municipal Corporations, often leading to conflict and contestations between the various urban local bodies.

6 The jurisdiction of BDA is limited to the city of Bangalore. However, that of BESCOM extends beyond the Bangalore urban region.

7 The Vision 2023 document is meant to function as a blueprint for revitalising the state's economy: its aims include increasing the per capita income of the state's residents to match that of Upper Middle Income countries, achieving a higher Human Development Index (HDI) to match those of developed countries, and providing high-quality infrastructure across the state (Government of Tamil Nadu et al., 2012).

8 The seven focal areas included: sustainable agriculture, water resources, forests and biodiversity, coastal area management, energy efficiency (including renewable energy and the solar mission), sustainable habitat and knowledge management.

9 This programme has been funded by the German Federal Ministry for Economic Cooperation and Development.

10 Until June 2016, Chennai had a directly elected mayor. In June 2016, the Tamil Nadu state government introduced a bill that would empower councilors to elect the mayor rather than having the mayor be directly elected by city residents. (B. Kolappan, 2016)

11 Surat was one of three Indian cities chosen to be part of the Asian Cities Climate Change Resilience Network (ACCCRN) funded by the Rockefeller Foundation. The other two cities were Gorakhpur and Indore.

References

Bangalore Mirror Bureau. (2016). NGT directs Mantri to pay Rs 117 cr, demolish Agara encroachment. *Bangalore Mirror*. Retrieved from http://bangaloremirror.indiatimes.com/bangalore/cover-story/NGT-directs-Mantri-to-pay-Rs-117-cr-demolish-Agara-encroachment/articleshow/52114213.cms?prtpage=1

Bulkeley, H. (2010). Cities and the Governing of Climate Change. *Annual Review of Environment and Resources*, 35, 229–253.

Byravan, S., and Rajan, S.C. (2012). *An Evaluation of India's National Action Plan on Climate Change*. Chennai, India: IFMR – Centre for Development Finance and IIT.

Census of India. (2011). *Provisional Population Totals, Census of India, 2011: Urban Agglomerations/Cities Having Population 1 Lakh and Above*. New Delhi, India: Census of India.

Dubash, N.K. (ed.). (2012). *A Handbook of Climate Change in India: Development, Politics, and Governance*. Oxon, UK and New York: Earthscan.

Dubash, N.K., and Jogesh, A. (2014). From margins to mainstream? State climate change planning in India. *Economic & Political Weekly*, 49(48), 86–95, Sameeksha Trust.

Dubash, N.K., and Joseph, N.B. (2015). The Institutionalisation of Climate Policy in India: Designing a Development-Focused, Co-Benefits Based Approach. *Centre for Policy Research, Climate Initiative, Working Paper.*

Ghertner, D.A. (2011). Gentrifying the state, gentrifying participation: Elite governance programs in Delhi. *International Journal of Urban and Regional Research*, 35(3), 504–532, Blackwell Publishing Ltd.

Government of Tamil Nadu (2014). *State Action Plan on Climate Change of Tamil Nadu State* (Final Draft).

Government of Tamil Nadu, Asian Development Bank (ADB) & iMacs (2012). *Vision 2023: Strategic Plan for Infrastructure Development in Tamil Nadu.*

IIHS, Development and Planning Unit (DPU) UCL & Atkins. (2013). *Urban Diagnostic for Bangalore: Future Proofing Indian Cities*. Bangalore, India: IIHS, DPU (UCL), Atkins.

Institute of Development Studies (IDS). (2007). *Governance Screening for Urban Climate Change Resilience-Building and Adaptation Strategies in Asia: Assessment of Chennai City*. India: IDS Climate Change and Disasters Group, University of Sussex.

The International Bank for Reconstruction and Development/The World Bank (2010). *Cities and Climate Change: An Urgent Agenda*. Washington, DC: The International Bank for Reconstruction and Development/The World Bank.

Janakarajan, S. (2016). The Cauvery water dispute. *Economic & Political Weekly*, 51(41), 10–15, Sameeksha Trust.

Jogesh, A., and Dubash, N.K. (2014). Mainstreaming climate change in state development planning: An analysis of Karnataka's Action Plan on Climate Change. *Centre for*

Policy Research (CPR), Climate Initiative, Working Paper. New Delhi: Centre for Policy Research (CPR).

Kolappan, B. (2016). The politics over mayoral election in Tamil Nadu. *The Hindu*. Retrieved from www.thehindu.com/news/cities/chennai/the-politics-over-mayoral-election-in-tamil-nadu/article8762161.ece

Kundu, D. (2011). Elite capture in participatory urban governance. *Economic & Political Weekly*, 46(10), 23–25.

Ministry of Environment and Forests. (2016). *State Action Plan on Climate Change*. New Delhi, India: Ministry of Environment and Forests, Government of India. [Online]. Retrieved from http://envfor.nic.in/ccd-sapcc

Mukhopadhyay, P., and Revi, A. (2009). Keeping India's economic engine going: Climate change and the urbanisation question. *Economic and Political Weekly*, 44(31), 59–70.

Nair, J. (2015). Indian urbanism and the terrain of the law. *Economic & Political Weekly*, 50(36), 54–63.

National Institute of Urban Affairs (N.I.U.A.). (2005). *Impact of the Constitution (74th Amendment) Act on the Working of Urban Local Bodies*. New Delhi: National Institute of Urban Affairs (NIUA).

Pasquini, L., Ziervogel, G., Cowling, R.M., and Shearing, C. (2015). What enables local governments to mainstream climate change adaptation? Lessons learned from two municipal case studies in the Western Cape, South Africa. *Climate and Development*, 7(1), 60–70.

Pinto, M. (2000). *Metropolitan City Governance in India*. Thousand Oaks, CA: Sage.

Prime Minister's Council on Climate Change. (2010). *National Action Plan on Climate Change*. New Delhi: Government of India.

Revi, A. (2008). Climate change risk: An adaptation and mitigation agenda for Indian cities. *Environment and Urbanization*, 20(1), 207–229.

Revi, A., and Sami, N. (2013). Social science research and global environmental change in India and South Asia. In: ISSC and UNESCO, *World Social Science Report 2013: Changing Global Environments*. Paris: OECD Publishing and UNESCO Publishing, 198–206.

Revi, A., Satterthwaite, D., Aragon, F., Corfee-Morlot, J., Kiunsi, R.B.R., Pelling, M., . . . Pahwa Gajjar, S. (2014). *Chapter 8: Urban Areas, Working Group II Report to the IPCC's Fifth Assessment Report*. Intergovernmental Panel on Climate Change (IPCC)

Sami, N. (2013). Power to the people? A study of Bangalore's urban task forces. In: Shatkin, G. (ed.) *Contesting the Indian City: Global Visions and the Politics of the Local*. West Sussex, UK: Wiley Blackwell, 121–144.

Sami, N. (2017). Learning by doing: Urban planning in Bangalore. In: Vidyarthi, S., Mathur, S. & Agrawal, S. (eds.) *Understanding India's New Approach to Spatial Planning and Development: A Salient Shift?* Delhi, India: Oxford University Press, 90–113.

Sharma, D., and Tomar, S. (2010). Mainstreaming climate change adaptation in Indian cities. *Environment and Urbanization*, 22(2), 451–465.

Stone Jr., B. (2012). *The City and the Coming Climate: Climate Change in the Places We Live*. New York: Cambridge University Press.

Surat Climate Change Trust (2015). *Surat Climate Change Trust* [Online].

Tanner, T., Mitchell, T., Polack, E., and Guenther, B. (2009). Urban governance for adaptation: Assessing climate change resilience in ten Asian cities. *IDS Working Papers*, 315, 1–47.

United Nations. (2014). *World Urbanization Prospects: The 2014 Revision*. New York: United Nations, Department of Economic and Social Affairs, Population Division.

Vijayabaskar, M. (2010). Saving agricultural labour from agriculture: SEZs and politics of silence in Tamil Nadu. *Economic and Political Weekly*, 45(6), 36–43.

Weinstein, L. (2009). *Redeveloping Dharavi: Toward a Political Economy of Slums and Slum Redevelopment in Globalizing Mumbai.* Doctor of Philosophy, Chicago: The University of Chicago.

Weinstein, L., Sami, N., and Shatkin, G. (2013). Contested developments: Enduring legacies and emergent political actors in contemporary urban India. In: Shatkin, G. (ed.) *Contesting the Indian City: Global Visions and the Politics of the Local.* West Sussex, UK: Wiley Blackwell, 39–64.

7 Climate change adaptation, city competitiveness and urban planning in the city of Karlstad, Sweden

Mikael Granberg and Lars Nyberg

Introduction

Sweden is often considered a pioneer in environmental governance (Lidskog and Elander 2012) and combining nationally high ambitions with strong municipalities.[1] Sweden provides an interesting context for case studies of local environmental politics and for climate change action (Granberg and Elander 2007; Granberg et al. 2016). The Swedish government system is an integrative central-local government system, where the municipalities have substantial financial, constitutional, legal, political and professional resources. With its extensive municipal output of welfare functions, Sweden has been labelled a decentralised welfare state (Sharpe 1988; Granberg 2008).

Accordingly, Swedish municipalities are crucial actors within the government system of the welfare state with a responsibility for local and regional matters of public interest on the principle of local self-government and may levy tax (including income tax) for the management of their affairs. In addition, land-use planning, including water areas, within the municipal boundaries is a municipal responsibility since 1948 (SFS 2010, p. 900). This is generally referred to as the municipal "planning monopoly". In principle, the authority to adopt plans and to decide on the implementation of plans rests with the municipality. However, municipal planning suggestions or decisions can be overturned if strong national interests are at risk. Hence, the state provides frameworks for the municipal planning through national objectives and by identifying claims of national interest.

Accordingly, national legal and administrative frameworks guide planning implemented by the municipalities through planning processes that produce guidance and decisions on land use and the allocation of building permits. Following the guidance of the comprehensive plan, municipalities produce what is known as a detailed development plan that provides a more operational level of planning and is a binding executive planning instrument. Adding climate change into this mix opens a potential for some temporal tension. While statutory comprehensive plans set out policies and proposals for 10 to 20 years ahead, the implications of climate change and connected risks have a much longer time horizons. Swedish planning legislation includes a mandatory requirement since January 2008, with further reinforcement in revised legislation enacted in 2011, that municipalities shall take flood risks, erosion and landslides into consideration in comprehensive and detailed development planning and in decisions over building permits (SFS 2010, p. 900, SKL 2011; Länsstyrelserna 2012).

Swedish municipalities also have a central role in handling environmental issues at large, including sewerage, waste-treatment, recycling, green public procurement, green consumption and green accounts (Granberg and Elander 2007; Lidskog and Elander 2012; Granberg forthcoming). Hence, Swedish municipalities have considerable resources but also extensive responsibilities within the frame of the welfare state. Accordingly, the policy agenda in Swedish municipalities consists of a number of high status issues, e.g., unemployment, segregation and city growth, etc., competing with climate change issues, illustrating an always present and pressing challenge for municipal policymakers: prioritising among policy issues and agendas.

Hence, the framing of, and competition between, policy issues is important when analysing policy processes (Mettler and Sorelle 2014) as framing is limiting the scope for what can be included in a policy. Policy scholar Hugh Heclo has stated this clearly when saying that policy is defined not only by what is included on the policy agenda, but also what is downplayed or excluded (1972, p. 85). Accordingly, in order to enable an understanding of the preconditions of climate change adaptation, the handling of climate risks on the planning agenda and efforts to reduce these risks in a specific geographical setting, we need to look closer at the priorities and politics of the wider policy context (cf. Granberg et al. 2016).

Accordingly, climate change adaptation does not take place in isolation from other policy items and decisions but are enmeshed with cultural change, demographic development and the globalising flows of capital (Adger et al. 2005, Moloney and Fünfgeld 2015). Following this perspective, when looking at local climate change action, a central question is what can we expect from the municipalities in terms of activity around climate change adaptation planning and related issues on the policy agenda and in the actual planning documents? Put in another way, what happens when climate change risks, vulnerabilities and needed adaptation measures confront other prioritised goals on the local policy agenda? In the case studied here, we find even in the environmentally progressive Swedish context in which the studied city is sometimes considered a forerunner, that economic growth imperatives continue to dominate and also frame climate risks understanding, potentially leading to constraints on the capability for local climate change action. Accordingly, this chapter looks closer at how municipal climate change responses are mediated through dominant policy priorities shaping city development goals and processes.

We do this through a case study of the city of Karlstad, Sweden. The case study focuses on the large-scale, long-term and ongoing projects of city redevelopment (1990s to present) with a special focus on the city's inner harbour area. The long-term focus makes it possible to study how climate change has emerged and been interpreted over time in plans and policy documents but also how this connects to other policy priorities and to the redevelopment of waterfront locations in the city. Accordingly, the material consists of the output of planning and city redevelopment policy: municipal planning documents, political visions for the city and public policy statements and documents. Comprehensive plans from the period 1997 to 2012, detailed plans, risk assessments, the flood risk management program and the vision for the city are used in the study (for a different analytical perspective on the same empirical material, see Granberg et al. 2016).

Studying these documents, we can observe that Karlstad is progressive and thorough in climate change adaptation action (cf. Davoudi et al. 2009, Hurlimann and March 2012; Storbjörk and Hjerpe 2013) – at least when it comes to more narrow adaptation perspectives focusing on increased water flows (compare with experiences from Denmark in, Hedensten Lund et al. 2012). We also clearly see how competing policy agendas – such as city competitiveness – are given high priority. It is evident that the political actors, political parties and individual politicians have played a central role in driving, and giving priority to, the economic growth competitiveness agenda interpreting the inner harbour as an asset that needs to be exploited (Granberg et al. 2016, Karlstad kommun 1998, 2008, 2010a). The political majority, a coalition of the four right wing parties and the green party that has been governing since 2010, has shown clear political leadership pointing out the direction. A focus on collaboration with other growth actors (businesses), including some large multinational construction and developing companies, has also been instrumental in driving the agenda. This clear political will has also given the planning process a frame that has not, at least not with any significant impact, included the option of 'no development' on, or 'retreat' from, the waterfront based on a prioritisation of climate risks.

A widespread international trend, often connected to post-industrialisation (cf. Ferm and Jones 2016), is the redevelopment of city areas in close proximity to waterfronts in efforts to enhance the aesthetic attractiveness of a city and, in turn, stimulate economic growth (Granberg 2008; Desfor et al. 2011; Galland and Hansen 2012; Boland 2013) along the lines of the long-lasting and powerful idea of the city as a growth machine (Molotch 1976). Hence, waterfront redevelopment often evolves through market-driven planning, focusing on the transformation of derelict areas into growth-promoting assets (Galland and Hansen 2012).

In the studied municipal documents from Karlstad, we can discern similar tendencies and perceptions as the development of the waterfront is framed as an opportunity to counteract negative impacts connected to economic globalisation through increased city attractiveness creating, and increasing, city competitiveness and facilitating economic growth. Simultaneously, the waterfront redevelopment drives the city´s innovative climate action profile as it is being proactive in adaptive planning and housing development in the redevelopment of the city's post-industrial harbour. Accordingly, construction and innovative preventive measures explicitly addressing climate change adaptation in design and development are facilitating opportunities for investment from developers in flood-prone areas, creating a potential for economic growth.

At the same time, however, the waterfront redevelopment can be interpreted as risk producing in the sense that municipal policies, plans and actions facilitating waterfront utilisation for mixed-use buildings augments vulnerability to flood risks and potentially increases future demands for climate change adaptation measures in the waterfront areas under redevelopment. This puts the spotlight on the complexity and duality of municipal action, showing that municipalities can be both part of climate change solutions but also instrumental in exacerbating problems, vulnerabilities and challenges related to climate change (cf Storbjörk 2007, van den Berg and Coenen 2012).

Below we will first give an overview of the city case study context and of the studied documents. We will look at the consultation process from a power and politics perspective and examine the tensions that are emerging between climate change adaptation and the goal of city competitiveness. Finally, we offer some concluding reflections on how these tensions play out in urban planning processes and the potential consequences of this for future climate risk management and adaptation in the city.

The city of Karlstad

Karlstad is a small but expanding city (90,086 in 2016 with a yearly increase of approx. 1 percent) situated on northern Europe's largest lake and river delta between Lake Vänern and the Klarälven River. The city has suffered a number of floods, notably in 1916, 1987, 1995 and 2000/2001 (Blumenthal 2010; MSB 2011a; Moberg 2011). In 2000/2001, heavy rains caused gradually increasing water levels in the lake, leading to a flood that lasted for 6 months covering extensive parts of the city, creating considerable damage, disruption and costs (Blumenthal 2010). During the period the municipality built five kilometres of temporary levees in the city protecting, among other areas, the inner harbour, several roads were closed, the municipal water treatment plant was threatened, storm drains was closed to prevent flooding and about 100 properties were threatened by the flood and several buildings were damaged.

Karlstad has been highlighted by the State Commission on Climate and Vulnerability as one of the Swedish cities with the highest risk of flood impacts

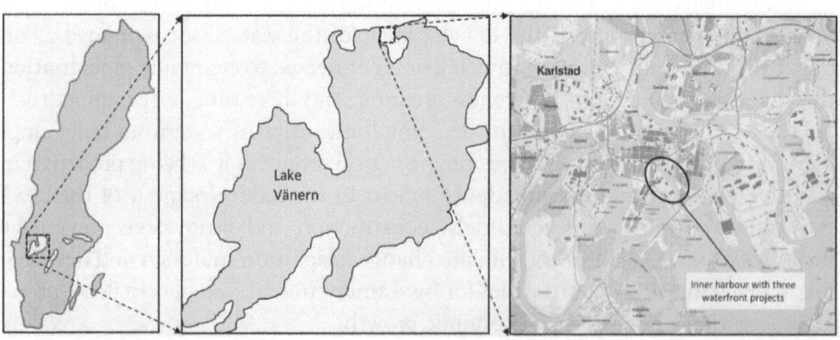

Figure 7.1 Map of Sweden, Lake Vänern and Karlstad with the inner harbour

related to climate change (SOU 2007:60). This has also been reinforced by the EU that has pinpointed Karlstad as a city facing extensive flood risks. At the same time, the United Nations (UNISDR) has identified the city as a 'Resilient City' (MSB 2011b). Karlstad also is one of the Swedish cities that is most active when it comes to waterfront development through the redevelopment of the post-industrial inner harbour (Hakkarianen 2013; Granberg et al. 2016).

The vision: Karlstad 100,000

The vision, produced and published by the municipality, of the city of Karlstad is to reach the goal of 100,000 inhabitants within the period 2030–2035 (Karlstad kommun 1998, 2008, 2010a). The aim is to attract new inhabitants by using and reinforcing the inherent qualities of the city, and a quality of central importance is the closeness to the lake and the river. Central components in this approach are the focus on competitiveness, the need for strong political leadership and vision from the municipality, a focus on business actors and growth coalitions and the need to create the image of an attractive city through redevelopment of the inner harbour.

In this strive for a competitive edge facilitating population growth, the municipality has participated in the project Talent Attraction Management in the Nordic Regions and Cities run by the consultancy company Tendensor (Nisser and Frisk 2013). The aim of the project was to "illustrate how public and private actors can build a successful collaboration for Talent Attraction Management . . . to answer a number of key questions, such as: What tools and strategies have worked well for developing a city or region's attractiveness for talents?"[2] The logics of this approach is clearly translated into the vision discussed above and into the comprehensive plan of 2012 that states that the municipality's primary aim is to attract two main categories of new inhabitants: students and 'people with competitive competencies' (Karlstad kommun 2012, p. 8). The basic idea is that municipal and private investments in city redevelopment will increase the attraction for educated people with what is perceived as vital competencies that, in turn, strengthens the city's competitiveness.[3] A basic logic is that the attraction of a well-educated population will enable investment from, and localisation of, companies in the city.

As already touched upon above, a central resource (perhaps *the* central resource) in this scheme is the city's location on the river and the lake. The perception from the municipal leadership is that the potential of the area in the inner harbour to contribute to the competitiveness cannot be ignored. Accordingly, and as we will see in the planning documents, presented below, city redevelopment since the late 2000s focuses on the inner harbour with the clear aim to open the city towards the lake using the waterfront in its efforts to make the city more attractive (Karlstad kommun 2013, p. 5). As we will see below, the idea of creating an attractive waterfront area underpinning city competitiveness, as formulated in the vision Karlstad 100,000, has been central in guiding planning and city redevelopment.

Case study planning context: the expanding city and the inner harbour

Planning 1997–2012

Three generations of comprehensive plans will be discussed here, namely the plans from 1997, 2006 and the contemporary plan from 2012. The comprehensive plans provide a general picture of the municipality's perception of the future trajectory of the city as a whole and for the areas where detailed plans have been developed. We will also discuss the detailed development plans for specific localities: Barkassen (Karlstad kommun 2000), Tyggårdsviken (Sweco 2008) and Kanoten (Karlstad kommun 2013). All three plans include redevelopment of inner harbour areas.[4]

The comprehensive plan of 1997 came about in the context of an economic recession and displays a cautious incremental approach towards the development of the city (Karlstad kommun 1997). Despite this, waterfront housing was included in the plan but not clearly framed as contributing to competitiveness as in later plans. The discussion of flood risks was very limited and encompasses one half-page. The plan recommended that facilities that can be disturbed by floods should *not* be located within areas inundated by the 1995 spring flood. If a building or infrastructure development takes place within such an area, measures have to be taken to minimise damage in case of floods.

The following plan focusing on the inner harbour area was the detailed development plan for the area called Barkassen (Karlstad kommun 2000). The plan also had limited references to attractiveness and to the inner harbour's potential to contribute to city competitiveness. The inner harbour area was previously zoned for industrial purposes, but in the plan, this was changed to housing and commercial purposes, unlocking the potential for a more aggressive redevelopment of the area. In the Barkassen plan, no flood risk related to the lake or the river is discussed.

The comprehensive plan of 2006 came about in a very different policy and city context than the plan of 1997 (Karlstad kommun 2006). The period was characterised by optimism, economic growth and challenges of city expansion was high on the policy agenda. This clearly shaped the view of the future development of the city that focused on the need for expansion in terms of new housing and increased competitiveness. The focus on attractiveness and city competitiveness is guided by the vision Karlstad 100,000 (Karlstad kommun 1998) that envisioned a growth rate of 500 people per year over the coming 30 years. Realising the vision put considerable pressure on the ability to plan, build new areas for housing and services and the strategic, not just the spatial, dimension of the comprehensive plan was stressed in the document. The planning focus shifted towards areas adjacent to the lake and the river. Despite the increased focus on waterfront locations, the discussion on the relevance of flood risks was still modest and encompasses just one half-page. The plan concluded that there was a risk of considerable infrastructure damage with high economic costs and environmental risks due to vulnerabilities to floods. The comprehensive plan also included a statement from the County Administrative Board (Länsstyrelsen) that considerations of flood risks should be further elaborated and included in the planning of housing within potential flood risk areas.

The detailed development plan for Tyggårdsviken (Sweco 2008)[5] focused on industrial areas in the inner harbour that gradually was being abandoned, thereby opening opportunities for the redevelopment of the inner harbour into an integrated city environment. The flood risks was presented under a dedicated heading and risk management was introduced supported by a discussion in one of the reports from the State Commission on Climate and Vulnerability (SOU 2007, p. 60). The plan discussed the dimensioned level[6] and the flood levels with a 100-year return period. A proposed solution was to build storage facilities, shops and restaurants at ground level as a buffer and for housing starting at the second level of the building. Vital societal functions, the plan stated, must be safeguarded to accommodate a 100-year return period.

The comprehensive plan of 2012 (Karlstad kommun 2012), the plan currently in force, states that the city should be developed in close proximity to the water, as this is a central aspect contributing to attractiveness that, in turn, strengthens city competitiveness. Again, the guiding document is the vision of Karlstad 100,000 (Karlstad kommun 2008). Flood risks are acknowledged but have to be handled, adjusted and interpreted within the framework of the vision with its aim of increased competitiveness and city growth.

The discussion on flood risks in the plan is much more wide-ranging in comparison to earlier planning documents, and flood risks are discussed under several headings in the plan. Under the heading 'Flood risks', closeness to water is identified as an important attractive aspect contributing to the city's competitive edge and, as such, an important resource for the city and the municipality. At the same time, it is concluded that the city's location in a delta landscape entails vulnerabilities related to risks of high water levels and floods.

This is interpreted as a double challenge for planning, expansion and redevelopment of the city, as waterfront development needs to consider flood risks. It is stated that new and increased knowledge on climate change and associated risks has changed the planning prerequisites for the city and that this has to be mirrored in the planning and future development of the city. Detailed plans within risk areas should include a risk analysis and suggest adaptive measures to be included reducing risk and vulnerability. In some areas, building will only be allowed if preventive adaptive measures regarding flood risk are implemented.

The detailed development for the area Kanoten primarily focuses on areas that have been developed through landfills in the lake, creating the ground for harbour development during the 1850s and 1960s (Karlstad kommun 2013). The plan includes further landfills to enable the expansion of exploitable land. The argument is that this answers to a strong common interest of city growth that entails an increased need for housing in the expanding city's waterfront. A guiding statement in the plan is that "Karlstad is an expanding city with an explicit political vision . . . with high ambitions of a viable and attractive city. . . expanding the city centre in the area towards lake Vänern" (Karlstad kommun 2013, p. 5).

Under the heading 'Economic consequences', this rationale is explained in two sentences simply stating that this redevelopment is economically beneficial for the economy of the municipality in several ways and that infrastructure is already in place (Karlstad kommun 2013, p. 20). In the plan, the redevelopment is described as directly being connected to Karlstad's strategic vision Karlstad 100,000, revised in 2010, that stresses the importance of creating a sustainable and attractive city able to handle an expansion to the ultimate goal of 100,000 inhabitants (Karlstad kommun 2010a). The plan makes clear that the development of this area is integrated in the broader competitive policy agenda and that the inner harbour is a vital resource in the realisation of this policy.

New buildings must, as stated in the municipal flood risk management program (Karlstad kommun 2010b), accommodate the flood levels of a 200-year return period. In addition, a separate study of the flood risks of the area has been done, and it is concluded that considerable parts of the area will be flooded at the contemporary flood levels of a 100-year return period. The potential possibility of retreat is touched upon very briefly but is not considered as a viable option in the plan that is driven by the logic of city growth (Karlstad kommun 2013, p. 25). Instead, two potential adaptive measures are formulated: landscape adjusted in-flow barriers and a lock on the inlet to the inner harbour. In addition to these measures, guidelines are also put in place that aim at protecting the vital functions of the buildings (electricity, water, etc.), with the aim of enabling functionality for the coming 100 years.

Political steering and power in the planning process

In the Swedish planning system, a decision on a detailed development plan is following a mandatory consultative process that should facilitate inputs, and even influence, from the county administrative board that is responsible for the supervision of the planning and building administration, the cadastral authority, other national authorities protecting claimed national interest (for instance, the National Board of Housing, Building and Planning (Boverket) that is responsible for the general supervision of the planning and building administration within the country), other departments of the municipal organisation, local stakeholders (business owners, property and real-estate owners, affected organisations, etc.) and individual citizens (SFS 2010, p. 900, chapter 5, section 20–28).

The purpose of the consultative process is to ensure that organisations and individuals affected by the plan proposal are offered opportunities to take part in consultation, raising concerns and giving constructive input. The overarching purpose is to improve planning decisions and to enable insight and influence from external actors. In the documents used by the municipality on a proposed detailed development plan, the reasons for the plan shall be put forth and after the conclusion of the consultation process the input from other actors should be reported. In the following sections we will, building on the consultation process reports from the planning processes of Barkassen, 2000, and with a special focus on Tyggårdsviken, 2008, and Kanoten, 2011, discuss issues of politics and power in the planning processes.

The issue of retreat from the waterfront in terms of no, or at least less, development has been raised in the planning process for the inner harbour area (Karlstad kommun 2009, 2010c). These voices have, however, had no impact on planning decisions and implementation. The critical inputs arguing for a retreat option focused on high water levels, floods, protection of the coastline, sensitive ecology in the area (partly a Natura 2000 area) and issues of ground-stability in areas developed through landfills in the lake or that includes further landfills to enable the expansion of exploitable land (Kanoten) (cf. Karlstad kommun 2009, p. 1, 2010c, p. 1). The County Administrative Board, among others, stated that the proposed measures to handle flood risk are not enough and that further measures are needed (Karlstad kommun 2010c, p. 3, 2013). In addition, the board argued that exploitation of the inner harbour area needs to seriously take potential impacts of a changing climate into consideration, reducing vulnerabilities (Karlstad kommun 2009, p. 4). This had some impact in the plan that stated that when developing the inner harbour, flood risks have to be given a prominent position in risk analysis and planning (Karlstad kommun 2010c). Changes and supplements to the plans influenced by the consultations has included supplemental data and elaboration of the map of the area, a geotechnical study of the ground stability and of the ground permeability in some parts of the areas subject to the plans (Karlstad kommun 2009, p. 1).

Input from construction companies, and also from some of the land and property owners, concerning the detailed development plans for the inner harbour area Kanoten agree with the standpoint that exploitation of the area is necessary, and one of the large multinational construction companies sought more aggressive exploitation of the area, allowing for higher building levels than what is proposed in the plan (Karlstad kommun 2010c, p. 21).

Arguments against the proposed plans for the inner harbour area are turned away with the claim that "that they are not strong enough to motivate that the area is not developed" (Karlstad kommun 2010c, p. 1). From the planning documents studied, it is evident that the steadfast political line has been that the inner harbour is a central asset that has to be exploited in ways that promotes competitiveness and that potential vulnerabilities connected to the exposure to flood risks has to be handled within the overarching objectives of this policy context (Karlstad kommun 1998, 2008, 2010a). The development of the area is formulated as a political vision addressing a strong public interest (Karlstad kommun 2013, p. 5, 10). Accordingly, the option of retreat has not at any point, at least not after 2000, been given any serious consideration by the political majority of the municipality.

Municipal climate change adaptation action

From the studied plans, we can observe that Karlstad is progressive, thorough and innovative in climate change adaptation action. The awareness of the need for handling flood risks and reduce vulnerabilities has had a considerable development over the years covered by this study. Flood risks have gradually risen in importance, as have the quality of the analysis and the discussion of the issues in the plans. It is clear that the spring flood of 1995, the lake flood in 2000–2001 and the findings by the State Commission on Climate and Vulnerability (SOU 2007, p. 60) includes important lessons for the municipality. This has led the municipality to increase its climate adaptation and risk reduction activity both in the inner harbour but also in the city as a whole. Below we will go through some of the municipal climate action.

The municipal flood risk management program: The work on the program started in 2006 through a decision in the municipal council (Karlstad kommun 2010b; Moberg 2011). One of the arguments for initiating this work was that Karlstad had been highlighted by the State Commission on Climate and Vulnerability (SOU 2007, p. 60) as one of the Swedish cities with the highest risk of flood impacts (Karlstad kommun 2010b, p. 6). The knowledge and analysis produced by the State Commission were also important inputs into the program. A number of studies and investigations into flood risks in the Karlstad region followed and the program was formally launched in 2010, again through a decision in the municipal council (Karlstad kommun 2010b). The flood risk management program is comprehensive and covers both contingency and prevention (Karlstad kommun 2010b).

The overarching objective of the program is to facilitate and sustain short- and long-term preparedness for facing floods risk and to minimise damage when it occurs (Karlstad kommun 2010b, p. 7). This will be done by establishing guidelines for planning and construction and create criteria for the protection of vital societal functions, to take necessary operative, technological and planning actions, and to openly inform and communicate flood risks both internally and externally. The objectives in the program are followed up and evaluated annually, and the municipal flood plains manager (presented and discussed in more detail below) compiles the results and reports to the central municipal administration that, in turn, reports to the political decision makers (Karlstad kommun 2010b, p. 6).

Collaboration with stakeholders both internally and externally: The municipality states that collaboration is a central issue with great potential in risk reduction, both in prevention and emergency issues, and that it is important that this also guides the development of the program. In the development of the municipal flood risk management program, a great number of actors were involved from municipal departments, emergency services, industry and businesses, researchers and representative for state authorities (Karlstad kommun 2010b).

The flood plains manager: In 2007, as the first municipality in Sweden, the municipality employed a full-time flood plains manager (Moberg 2011). The main mission of the flood plains manager was to build knowledge and increase capacity and the level of ambition in the municipality's flood risk management processes. This entailed initiating action in several municipal departments that had activities with the potential to be impacted by, or mitigate vulnerabilities to, flood risks. Several actions were consequently initiated, including:

- two surveys by the Swedish Hydrological and Meteorological Institute (SMHI),
- airborne laser scanning of flood-prone areas,

- hydrodynamic modelling,
- inclusion of flood maps in the municipal GIS-system,
- information activities towards politicians in the municipality, and,
- the implementation of a water level and flow measurement program for the river.

Communication and outreach: Aside from the technological measures, the municipality considers communication and outreach as important activities aiming at strengthening the city's capacity for adaptation (Karlstad kommun, undated). The municipality's outreach program entails flood maps being available on the Internet, brochures informing on flood risks, articles discussing and debating flood issues in news media, information to industry and business of flood risks and mitigating efforts and collaboration with schools and with universities. The issue of communication of flood risks to the public, private actors and organisations was also highlighted in the work on the flood risk management program and reinforced by the work of the flood plains manager (Moberg, 2011).

Flood walks: The Centre for Climate and Safety at Karlstad University had started developing the concept of a flood walk directed to the public, and the municipality joined this endeavour (Johansson et al. 2013). The basic idea is that flood risks are a moving target, and it is important to offer opportunities of recurrent education and development of well-adapted didactics for increasing the ability for risk reduction through the transfer of risk knowledge. Flood risk is produced at the interface of natural and social systems, and it is important to address and illustrate the interdependence between these systems. In the flood walks this is done by using risk areas as a field-based "classroom" providing learning through multiple aspects in a communication between the public and researchers/experts, giving a major input about flood hazards and societal consequences as well as strategies to reduce risks and giving the possibility to examine examples of events and risks on-site providing hands-on information in the field, using different types of knowledge (expert, experience based, etc.) and all senses.

Physical measures: In 2015, the municipality began construction of a moveable flood barrier at the inlet from the lake into the inner harbour. The flood barrier is built in connection to a road bridge and also includes pumping facilities to prevent flooding from heavy rains when the flood barriers are in place. The flood barriers are in place since mid-2016.

Reflections on climate change adaptation, climate risk management and city competitiveness

Analysing the planning documents presented above makes it clear that there is a growing awareness of the double challenge for planning and city redevelopment as the city develops waterfront housing that needs to consider flood risks in the strive for city competitiveness. An interesting observation is how the two issues of climate-related flood risks and city competitiveness develops over time in the studied planning documents.

The analysis of the planning documents shows that there was very little attention given to the issues of flood risks and competitiveness in the early plans.

Over the years studied, the two policy issues' importance has gradually increased. Particularly the floods of 1995 and 2000/2001 moved flood risks up on the urban planning agenda, and these issues were further reinforced by the identification of the Swedish Commission on Climate and Vulnerability (SOU 2007, p. 60) of Karlstad as one of the Swedish cities with the highest vulnerability to flood risks. This highlighted the importance of taking flood risks into account in urban planning at the city planning office. So, local extreme flood events, a national commission and its overview of flood events and future flood risks and connected to an increasing awareness, especially among municipal planners, of the potential future impacts of climate change have been important drivers for elevating the importance of flood risks on the city's urban policy and planning agenda.

The municipal growth strategy was put in force through a decision in the municipal council in 2010 and was a result of increasing political focus on the growth potential of the city (Karlstad kommun 2010a). It is clear that strong political leadership and a focus on collaboration with other growth actors (businesses) has had an important role in this development. The development was driven by the political majority, a coalition of the four right wing parties and the green party who have been in power since 2010. One important means for achieving growth has been the redevelopment of the inner harbour area that is thought to contribute to the creation of an attractive and competitive city. The increased growth of the city is, in turn, perceived as creating and further reinforcing competitiveness generating a continuous positive spiral of amplified competitiveness.

As stated already in the introduction, it is clear that the municipality of Karlstad is progressive and thorough in climate change adaptation action, at least when it comes to more narrow adaptation perspectives focusing on increased water flows, and we observe many activities often related to good practice climate change adaptation planning when it comes to physical measures but also when it comes to information, outreach and communication.

Hence, the city is both active and innovative in implementing policies and utilising actors to handle climate-related risks through the municipal flood risk management program, the full-time flood plain manager, etc., and through hands-on measures (barriers, levees and pumps) but also keen on spreading the knowledge to different actors in the local setting (collaboration, communication and outreach, flood walks, etc.). It is quite clear, however, that the focus on competitiveness and city growth is prioritised over the issue of climate risk vulnerabilities on the policy agenda. Competitiveness creates a frame for the understanding and implementation of climate risk management actions and measures for reducing vulnerabilities through climate change adaptation.

With a sharpened formulation, we can say that the municipal leadership perceive the city of Karlstad as part of an on-going 'place war' (Karlstad kommun 2008, 2010a, 2012; Nisser and Frisk 2013). To be successful in the 'war', to survive and flourish in the future, the city has to be attractive and develop its competitiveness. This perception impacts the policy agenda of urban planning and guides city redevelopment. Competitiveness emerges as the dominant goal in the local policy context, constraining the potential for other issues that do not clearly complement or reinforce the overarching aim of competitiveness. Accordingly, the measures to manage climate-related risks formulated in the planning documents are designed to accommodate flood risks in ways that do not conflict with city competitiveness and the increased risk awareness that we have observed in the planning documents does not, for instance, entail any suggestions of retreating from the waterfront.

Accordingly, the issues associated with adaptation to climate-related vulnerabilities, including limiting development in high-risk areas, are not able to compete with the dominant values of competitiveness in the city's redevelopment policy. It is clear, however, that the redevelopment of the waterfront will increase exposure to climate change–induced flood risks and that this will increase the need, and potentially the cost, for future climate change adaptation measures from the municipality and other actors. This seems to be part of a calculation entailing economic, social and ecological risks that results in the conclusion that climate-related risks are reasonable in the face of expected economic competitiveness gains from allowing development on the waterfront.

Even in a context like Sweden, known for its leadership in environmental policy including climate change risk management and planning (Lidskog and Elander 2012), the dominance of a growth imperative continues to outweigh the potential impacts of climate change risks. Hence, when the goal of economic growth dominates the city re-development agenda, we need to increase our understanding of how climate change risks are framed and how this has a clear impact in how these risks are defined, prioritised and managed. The framing of climate vulnerabilities by competitiveness is potentially dysfunctional as it runs the risk of not being able to accommodate effective climate change adaptation and risk management increasing the vulnerability of the city in both the short term and in the longer run. Such development will, ultimately, decrease city attractiveness and competitiveness. In this sense, municipalities are central actors that are both important for solutions but also clearly part of problems facing climate change (cf. Storbjörk 2007, van den Berg and Coenen 2012).

Notes

1 In this chapter, we use the term municipality when discussing Swedish local government.
2 The aims are taken from the projects website where a list of the participant cities and regions is also published (http://tendensor.com/news/tendensor-news/nordic-project-talent-attraction/).
3 These types of ideas are commonplace in city and regional development around the world and was, at least in part, introduced by Richard Florida (2002) and has been promoted by consultants as the way to increase competitiveness for cities and regions on a global market arena.
4 Quotes from the planning documents below are translated from Swedish by the authors of this chapter.
5 Sweco is a consulting firm focusing on engineering, environmental technology and architecture.
6 This level is used to dimension hydropower dams in Sweden and is also used as a low-probability level (lower than 1/10,000) for flood risk analysis.

References

Adger, N., Arnell, N.W., and Tompkins, E.L. (2005). Successful adaptation to climate change across scales. *Global Environmental Change*. Part A, 15, 77–86.

Blumenthal, B. (2010). *När Vänern svämmade över: Händelseutveckling och konsekvenser av översvämningen 2000/2001, Rapport 2010:1*. Karlstad: The Centre for Climate and Safety.

Boland, P. (2013). Sexing up the city in the international beauty contest: The performative nature of spatial planning and the fictive spectacle of place branding. *Town Planning Review*, 84(2), 251–274.

Davoudi, S., Crawford, J., and Mehmood, A. (eds.) (2009). *Planning for Climate Change: Strategies for Mitigation and Adaptation for Spatial Planners*. London: Earthscan.

Desfor, G., Laidley, J., Stevens, Q., and Schubert, D. (eds.) (2011). *Transforming Urban Waterfronts: Fixity and Flow*. London: Routledge.

Ferm, J., and Jones, E. (2016). Mixed-use 'regeneration' of employment land in the post-industrial city: Challenges and realities in London. *European Planning Studies*, 24(10), 1913–1936.

Florida, R. (2002). *The Rise of the Creative Class: And How It's Transforming Work, Leisure, Community and Everyday Life*. New York: Basic Books.

Galland, D., and Hansen, C.J. (2012). The roles of planning in waterfront redevelopment: From plan-led and market-driven styles to hybrid planning? *Planning Practice and Research*, 27(2), 203–225.

Granberg, M. (2008). Local governance in 'Swedish': Globalisation, local welfare government and beyond. *Local Government Studies*, 34 (3), 363–377.

Granberg, M. (forthcoming). Local government and climate change: Reaching low carbon futures between the state and the market? The case of Örebro, Sweden. In: Luque-Ayala, A., Bulkeley, H., and Marvin, S. (eds.) *Rethinking Urban Low Carbon Transitions*. London: Routledge.

Granberg, M., and Elander, I. (2007). Local governance and climate change: Reflections on the Swedish experience. *Local Environment*, 12(5), 537–548.

Granberg, M., Nyberg, L., and Modh L-E. (2016). Understanding the local policy context of risk management: Competitiveness and adaptation to climate risks in the city of Karlstad, Sweden. *Risk Management*, 18(1), 26–46.

Hakkarianen, H. (2013). Hållbar utveckling och översvämningsrisker – En analys av kommunala bedömningsgrunder vid stadsutveckling i vattennära områden. Master thesis in Societal Risk Management. Karlstad, Karlstad University.

Heclo, H.H. (1972). Policy analysis. *British Journal of Political Science*, 2(1), 83–108.

Hedensten Lund, D., Sehested, K., Hellesen, T., and Nelleman, V. (2012). Climate change adaptation in Denmark: Enhancement through collaboration and meta-governance? *Local Environment*, 17(6–7), 613–628.

Hurlimann, A.C., and March, A.P. (2012). The role of spatial planning in adapting to climate change. *Wiley Interdisciplinary Reviews: Climate Change*, 3(5), 477–488.

Johansson, M., Nyberg, L., Evers, M., and Hansson, M. (2013). Using education and social learning in capacity building – The IntECR concept. *Disaster Prevention and Management*, 22(1), 17–28.

Karlstad kommun (1997). Översiktsplan 1997. (Comprehensive plan).

Karlstad kommun (1998). Karlstad 100 000 (Visionary document).

Karlstad kommun (2000). Planbeskrivning Barkassen. (Detailed plan).

Karlstad kommun (2006). Översiktsplan 2006. (Comprehensive plan).

Karlstad kommun (2008). Livskvalitet Karlstad 100 000 (revised Visionary document).
Karlstad kommun (2009). Samrådsredogörelse: Detaljplan för Kv. Kanoten m.fl.
Karlstad kommun (2010a). Tillväxtstrategi (Growth strategy).
Karlstad kommun (2010b). Översvämningsprogrammet (Flood program).
Karlstad kommun (2010c). Samrådsredogörelse: Planprogram för Kv. Kanoten m.fl. (Consultation report).
Karlstad kommun (2012). Översiktsplan 2012. (Comprehensive plan).
Karlstad kommun (2013). Planbeskrivning Kanoten. (Detailed plan).
Karlstad kommun (undated). Levee – General hospital.
Länsstyrelserna (2012). *Klimatanpassning i fysisk planering – Vägledning från Länsstyrelserna* Stockholm: Länsstyrelsen.
Lidskog, R., and Elander, I. (2012). Ecological modernisation in practice? The case of sustainable development in Sweden. *Journal of Environmental Policy and Planning*, 14(4), 411–427.
Mettler, S., and Sorelle, M. (2014). Policy feedback theory. In: Sabatier, P.A. and Weible, C.M. (eds.) *Theories of the Policy Process*. 3rd ed. Boulder: Westview Press, 151–182.
Moberg, J-O. (2011). Karlstads översvämningsprogram- analys och utvärdering. In: Hodgin, S. and Mannheimer, J. (eds.) *Lokala riskhanteringsplaner mot översvämning – erfarenheter från Karlstad och Lidköping*, Rapport 2011:58. Göteborg: Länsstyrelsen i Västra Götaland, 23–32.
Moloney, S., and Fünfgeld, H. (2015). Emergent processes of adaptive capacity building: Local government climate change alliances and networks in Melbourne. *Urban Climate*, 14, 30–40.
Molotch, H. (1976). The city as a growth machine: Toward a political economy of place. *American Journal of Sociology*, 82(2), 309–332.
MSB (2011a). *Historiska översvämningar 1901–2010*. Karlstad: Myndigheten för samhällsskydd och beredskap (MSB).
MSB (2011b). *Identifiering av områden med betydande översvämningsrisk*. Karlstad: Myndigheten för samhällsskydd och beredskap (MSB).
Nisser, P-S., and Frisk, M. (2013). Karlstad – en attraktiv stad att leva i. 2013–11–05. Karlstad, *Nwt*. Retrieved from http://nwt.se/asikter/debatt/2013/11/05/3
SFS 2010:900. *The Planning and Building Act*.
Sharpe, L.J. (1988). The growth and decentralisation of the modern democratic state. *European Journal of Political Research*, 16(4), 365–380.
SKL (2011). *Kommunernas arbete med klimatanpassning*. Stockholm: Avdelningen för tillväxt och samhällsbyggnad, Sveriges Kommuner och Landsting (SKL).
SOU (2007).60 *Sweden Facing Climate Change – Threats and Opportunities*. Final report from the Swedish Commission on Climate and Vulnerability. Stockholm: Fritzes.
Storbjörk, S. (2007). Governing climate adaptation in the local arena: Challenges of risk management and planning in Sweden. *Local Environment*, 12(5), 457–469.
Storbjörk, S., and Hjerpe, M. (2013). 'Sometimes climate adaptation is politically correct': A case study of planners and politicians negotiating climate adaptation in waterfront spatial planning. *European Planning Studies*, 22(11), 2268–2286.
Sweco (2008). *Detaljplan Tyggårdsviken: Planbeskrivning och genomförandebeskrivning*. Stockholm: Sweco AB. (Detailed plan)
van den Berg, M., and Coenen, F. (2012). Integrating climate change adaptation into Dutch local policies and the role of contextual factors. *Local Environment*, 17(4), 441–460.

8 From engagement to empowerment

Climate change and resilience planning in Baltimore City

Kristin Baja and Mikael Granberg

Introduction

The societies we live in are getting ever-more complex and diverse and we argue, communities need to be conceptualised as "diverse social groups which may need to be treated differently" (Reeves 2005, p. 1). This also means that it is important to understand how historical dynamics have shaped the current social context through past interventions, disturbances and responses. (Wilkinson 2012; Resilience Alliance 2007, p. 22). Accordingly, this chapter touches on the historic, current and predicted stressors and shocks in Baltimore and describes how equity has been the backbone for engagement, outreach, coordination and plan development and implementation.

Hence, the historical context is critical for planners to understand and empathise with (Wilkinson 2012). Understanding what happened in the past – who has benefited, who has been impacted and where have funds been invested – is essential to consider and address from a racial equity standpoint. The history must be visible and acknowledged throughout the process. The difference between equity and equality is frequently misunderstood, particularly in city planning where historical policies and practices intentionally utilised race and class to segregate specific populations (cf Reeves 2005). Hence, understanding the difference between equity and equality is a key component. Equality is treating everyone the same. It makes the assumption that if all people are given the same resources and options, that they are being treated fairly. However, that only works if everyone has the same set of circumstances. Equity involves understanding the history and context and ensuring people receive the resources they need in proportion to what they have been denied historically.

A truly equitable process requires honesty, real talk and a true recognition of one's own privilege (Reeves 2005; Wilkinson 2012). It requires participants, especially white participants, to acknowledge their own advantage and internal biases. It also requires city government to openly discuss and address institutional racism and historic racism in planning (Pulido 2000). That process is hard and can be a difficult for all parties involved. A challenge for planners and stakeholders guiding the processes and projects listed in this chapter is to utilise history to make connections to present-day circumstances without participants feeling blamed or defensive. It is an ongoing progression that involves continuous learning, acknowledgement and growth.

In Baltimore City, racism resulting from historic policies and planning has led to disenfranchised communities and inequitable access to resources and opportunities. The city of Baltimore, Office of Sustainability, strives to directly address racism and classism by utilising equity as a lens for all climate change and resilience planning and implementation processes. Using equity to guide Baltimore's climate and resilience efforts requires developing a common understanding of racial inequity and moving forward on eradicating disparities and achieving more equitable resolutions. Directing efforts towards more vulnerable neighborhoods strengthens the city overall; it helps empower stakeholders, builds trust, ensures residents' basic needs are being met and promotes community cohesiveness (cf Pulido 2000). Programs for job training, health prevention, food access, financial literacy, addiction recovery and education coupled with considerations for natural and human-made shocks are all critical pieces of improving resilience. Challenges associated with this type of process are discussed later in this chapter.

Although resilience planning and implementation encompass efforts around both natural and man-made hazards (Davoudi 2012), the city has found that leading with weather-related hazards and climate change and framing all of that work through an equity lens has been the most effective way to both communicate risk and to incentivise action. Through research and historical data, impacts from hazards can easily be tied to human health issues (e.g., asthma and heat stroke due to poor air quality and high heat), economic issues (e.g., loss of jobs and/or revenue when businesses are forced to shut down) and environmental issues (e.g., water pollution from sewage overflow). As with many cities in the United States, these social, economic and environmental impacts will disproportionately affect communities of color and low-income residents. Communicating how climate change exacerbates existing concerns that relate to people's everyday lives, and demonstrating proactive solutions which both individuals and the city can take, is the approach Baltimore is utilising to become a more resilient city.

Since 2012, the Office of Sustainability along with members of the Commission on Sustainability and community leaders have been discussing the role of race, equity and inclusion in both planning and implementation of all department initiatives. In order to conduct effective community engagement and empower residents, direct acknowledgment of racism and its integration into processes is critical to successful outcomes. A primary goal of the Office of Sustainability is to institutionalise equity into core city government work plans and implementation efforts. As an agency within the Department of Planning, the Sustainability Office is responsible for a range of plans, projects and permits. The office is deeply committed to advancing equity in Baltimore as a means to improve the way the city functions as well as the lives of its residents. Baltimore's history of deliberate segregation has directly led to economic and social disparities prevalent throughout the city. These disparities will be further exacerbated by impacts from weather-related disasters and climate change. For low-income residents and communities of color, there is a need for increased capacity to overcome additional stresses that may result from a disaster.

Actions taken to mitigate shocks and better prepare residents for climate impacts will have the co-benefit of improving chronic stressors and increasing resilience. Resilience is a very broad term. Like sustainability, it can mean many different things to different people (Porter and Davoudi 2012; Davoudi 2012). Baltimore City defines resilience as the ability of the community to anticipate, accommodate and positively adapt to, or thrive, amidst changing climate conditions or hazard events, and enhance quality of life, reliable systems, economic vitality and conservation of resources for present and future generations. This definition demonstrates the city's commitment to proactively improve essential infrastructure, buildings and natural systems while enhancing community adaptive capacity. This definition also establishes that climate change is just one of the many stressors Baltimore City faces. Efforts to adapt and mitigate shocks from natural hazards will also help improve the city's ability to tackle other stressors and man-made shocks.

Background and city context

Baltimore City, the 29th-largest city in the United States, is located within the Northeast megalopolis, one of the most urbanised regions in the country. Founded in 1792, Baltimore is the largest independent city in the U.S. and is home to the one of the largest seaports in the mid-Atlantic region (Maryland State Archives 2017). Baltimore historically developed as a port city and has a strong connection with water. With 60 miles of waterfront, and four major watersheds, the port and waterfront remain extremely important assets for Baltimore, supplying many jobs and some of the city's strongest property tax base. The city's economic activity has largely centered on waterfront development and redevelopment, including industrial, commercial, recreational and residential uses.

The city's growth and economy are tied to its location at the top of the Chesapeake Bay, the nation's largest estuary, and its position as the largest inland port on the east coast. Today, in addition to shipping, the city's economy is heavily tied to healthcare and medicine, financial services, advanced manufacturing and maritime commerce (Baltimore Development Corporation 2017).

Baltimore's location along the Chesapeake Bay is ideal for accessibility and economic growth. However, it also makes the city more vulnerable to impacts from a changing climate. The stream valleys and harbor that are the foundation of the city are also the areas at most risk. They house the city's most historic communities, infrastructure assets and international port. In fact, the communities that have seen the greatest population growth in the past decade are all located along the city's shoreline and waterways.

Baltimore's population is a little over 620,000 residents with 64 percent black, 29 percent white, 4.2 percent Hispanic or Latino (USCB 2010). The city's 80 square miles encompass the most heavily developed area within the state of Maryland. Known as a city of neighborhoods, Baltimore contains over 250 self-identified neighborhoods made up of brick row houses, office centers and university campuses. Although the city is an urban center with tremendous assets, the city also faces significant challenges. Baltimore's unique combination of shocks and stresses cuts across social, economic and environmental factors.

Baltimore has an unfortunate history of deliberate segregation that ties directly into many of the economic and social challenges the city faces today. In 1910, an ordinance was passed that partitioned Baltimore into black blocks and white blocks. Although this ordinance was soon invalidated, the practice of segregation nevertheless persisted for decades afterward. In addition, many white neighborhoods also adopted restrictive covenants that barred sales to minorities. The Federal Housing Administration's openly racist system for loan approval from the 1930s to 1960s prevented African Americans in Baltimore from gaining wealth or benefits associated with homeownership.

The effects of these policies are still evident in Baltimore City today. According to 2010 Census data, Baltimore is one of the most segregated cities in the United States. The city has an overall higher poverty rate than most US cities, and its poorer areas are mainly African American. African American residents in Baltimore have higher unemployment rates, more children living in poverty, less access to goods and services and lower educational attainment than white populations within in the city.

After the riots in 1968, ignited by the assassination of Martin Luther King Jr., Baltimore experienced a massive decline in its population, resulting in a weakened tax base. Over time, this trend has impacted the city's ability to address infrastructure improvements and human and social services. Maintenance and upgrades to aging infrastructure are woefully behind schedule, leading to frequent urban karsts, water main breaks and infrastructure failures

Natural hazards and historic impacts

Baltimore is highly vulnerable to a range of natural hazards, including coastal storms, flooding, extreme heat and high winds. Over the last six years, severe weather in the form of torrential rains, heat waves, tornadoes, derechos, blizzards and hurricanes have impacted the city and residents with increased intensity. Accordingly, the impacts of climate change are already being felt in Baltimore City. Since 2010, Baltimore has endured repeated severe weather events from such varied sources as heavy precipitation, tidal flooding, extreme snow and ice storms, coastal storms, heatwaves and extreme windstorms, including a tornado in the Inner Harbor. This increase in natural disasters will create potential impacts that will undoubtedly affect the city's residents, businesses, infrastructure and natural systems, and threaten regionally significant assets.

Currently, flooding is one of the two most threatening hazards in Baltimore. The region has experienced a significant increase in the amount of heavy precipitation falling during large storm events. Since 1948, extreme downpours occur 30 percent more often nationwide (Environment Maryland Research & Policy Center 2012). In the northeast region (inclusive of Baltimore), an increase in the amount of precipitation falling in very heavy events has increased by 71 percent since 1958 (National Climate Assessment 2014). Such extreme events are expected to increase in frequency and intensity in the winter and spring months (NOOA 2013b).

Most of Baltimore City's floods have been the result of flash flooding from intense cloudbursts or localised flooding due to poor drainage and stormwater management. Many of these floods have caused significant damage to infrastructure and buildings and at times, led to loss of life. Baltimore City is home to one of the most flood-prone watersheds in the region due to a high percentage of impervious surface and rapid development without concern for natural systems. Despite the watershed being a persistent flood threat, several of the city's historic mill buildings have been converted into mixed-use for residential and businesses, increasing the risk of property damage and loss of life. The majority of these buildings are repetitive loss properties due to frequent flash flooding and substantial losses. However, due to development pressure and local politics, much of the data highlighting the vulnerability of these properties has been disregarded in favor of economic growth.

On April 30, 2014, nearly five inches of rain fell in 12 hours, leading to a 14-foot rise in stream levels (City of Baltimore 2016). This storm led to the collapse of a retaining wall onto the CSX railway and flash flooding that destroyed property and businesses. In August of the same year, six inches of rain fell in less than five hours, shutting down major interstates, tunnels and roads and requiring several emergency rescues. In July of 2016, persistent heavy rainfall from another major storm cell led to an 11- to 16-foot change in stream levels, carrying several cars downstream and flooding businesses. This trend is likely to continue into the foreseeable future.

Heavy precipitation is just one of the ways water negatively impacts Baltimore City. The Union of Concerned Scientists released a report in 2014 that projects Baltimore will see an increase in tidal flooding from 17 days per year to 227 days per year within the next 30 years (Spanger-Siegfried, Fitzpatrick and Dahl 2014). Accordingly, tidal flooding, known as nuisance flooding, is also a growing threat to the city's waterfront assets and economy. Sea levels in Baltimore have risen over 13 inches since 1900 (NOAA 2013a). The Chesapeake Bay region is still undergoing isostatic adjustment, in which the removal of pressure from glacial ice that melted from the last ice age causes sinking of the land (Larsen 1998). In the mid-Atlantic region, land movements associated with isostatic adjustment the melting of ice in polar regions, and groundwater withdrawal is leading to some of the highest rates of relative sea-level rise along the United States Atlantic Coast (Boon, Brubaker and Forrest 2010; Cronin 2013). Sea levels in the Chesapeake Bay are anticipated to increase by two feet within the next 35 years, nearly twice the average global rate (University of Maryland Center for Environmental Studies 2014). Coastal storm surges may be amplified by sea level rise, creating a greater threat. As extreme events, combined with a projected rise in sea level, impact the region, these effects will be spread over a much larger area of the city and threaten regionally significant utilities, such as sewage treatment plants and Baltimore's waste-to-energy facility. Despite obvious threats from water-related risks, the trend of exploitation of Baltimore's already densely developed waterfront continues.

In 2003, Hurricane Isabel hit the east coast of the United States. In Baltimore, the worst impacts resulted from the over seven-foot storm surge that inundated the historic waterfront district of Fells Point and the Inner Harbor, causing millions of dollars' worth of damage for homeowners and businesses. In 2012, Hurricane Sandy hit the northeast Atlantic Coast, bringing a nine-foot storm surge to areas in New York and New Jersey (USGS 2015). Due to its location at the top of the Chesapeake Bay, had the surge from Sandy hit Baltimore City in combination with freshwater flooding from rain, the impacts to structural systems, communities and the economy would have been tremendous.

Higher temperatures and increased frequency of heat waves are the other major threat to the city. Baltimore summers typically include both high temperatures and high humidity levels. During extreme heat events, temperatures in the city can be up to 10°F warmer than the surrounding counties. This urban heat-island effect is anticipated to intensify as global warming continues. According to NOAA, between 1981 and 2010, Baltimore experienced an average of 31 days a year when temperatures met or exceeded 90°F. In Baltimore, the number of days when temperatures reach or exceed 90°F are projected to increase to between 35 and 38 days by 2050, and to between 38 and 41 days a year by the end of the century (Walsh et al. 2014). By 2100, Maryland's average winter temperature are projected to increase by 2–6°F, and average summer temperatures are projected to increase by 3–9°F (Bradley et al. 2010). The resulting impacts from these shocks, even when it does not cause direct loss of life, can damage property, cause power outages, disproportionately impact vulnerable populations and interrupt businesses operations.

Stresses

In addition to a wide range of natural hazards, the city faces a number of social stresses, including a high incidence of crime, drug use and dependence and poverty. Many of these problems stem from a legacy of residential segregation drawn along race and class lines. Baltimore is one of the most segregated cities in the United States (United States Census Bureau 2010). The racial and social disparities in education levels, employment and access to services are severe. These chronic stresses facing the city serve to exacerbate the city's desperate need to protect its citizens from weather variability and climate change. There are seven major vulnerability factors influencing climate impacts: income, racism, language barriers, health conditions and disabilities, age, living conditions and occupation (Cooley et al. 2012; Jerrett et al. 2012; Morello-Frosch et al. 2008). In Baltimore City, those most vulnerable to the impacts are low income, elderly, young and/or people of color based on intentional segregation. Hence, Baltimore's significant social and economic issues will be further intensified with impacts from natural hazards and climate change.

Plan development

In 2000, President Clinton signed the Disaster Mitigation Act, which requires state, local and tribal governments to develop an All Hazard Mitigation Plan (AHMP) as a condition for receiving non-emergency disaster assistance. These plans must include an assessment of all historical hazards and vulnerability to those hazards, and identify mitigation strategies. Additionally, AHMPs must be updated every five years.

In 2012, Baltimore made a commitment to proactively address climate change utilising a unique planning approach that integrated climate adaptation with hazard mitigation efforts. Instead of following the traditional Federal Emergency Management Agency (FEMA) all hazards methodology, Baltimore focused on combining two historically disconnected processes into one comprehensive process and plan. An assessment of historic hazards was coalesced with evidence-based climate predications, which provided planners with a greater understanding of current and future risks. This proactive approach allowed the city to not only consider historic hazards, but also the opportunity to acknowledge existing climate change and integrate future climate considerations. The city also employed several other planning methodologies and engagement techniques to go above and beyond FEMA's minimum standards. This included utilising equity to frame the engagement process, ensuring that the voices of those most impacted, both historically and presently, were heard and incorporated.

Advisory committee

Baltimore City utilised a 40-person advisory committee of experts and stakeholders to inform and guide the planning process. The mayoral-approved committee consisted of climate scientists, local business owners, city agency directors, institutions, non-profits and residents. In addition, several subcommittees such as energy and infrastructure, the built environment, natural systems, public services, historic and cultural assets and the waterfront, were developed with additional experts and community members.

Input from the advisory committee along with feedback from residents and stakeholders led to the development of six goals and four main categories by which the plan was structured: infrastructure, buildings, natural systems and public services. These four categories contain a total of 50 strategies and 231 specific actions that currently guide Baltimore's climate and resiliency planning and implementation. Baltimore City felt implementation would be more successful if the plan's structure broadly mirrored those of city departments and implementing agencies in order to more effectively align planning and implementation with existing efforts.

In addition to altering the structure of the plan, the Office of Sustainability engaged in a cross-walking activity that identified overlaps between the plan's strategies and actions with existing local and state projects and plans such as Baltimore's Emergency Support Functions (ESFs), State of Maryland Adaptation and Hazard Mitigation Plans, and the City's Climate Action Plan. This helped prioritise actions for implementation, enhance existing projects, eliminate duplication of effort and integrate adaptation considerations into all existing efforts.

In October 2013, the Planning Commission and the Commission on Sustainability adopted the Disaster Preparedness Project and Plan (DP3) unanimously. Since adoption, the Office of Sustainability, with assistance from other agencies, community members and stakeholders, has focused on implementation of priority and short-term actions from the plan. Through innovative planning and implementation and strong collaborations, the city is making significant progress in increasing its adaptive capacity and reducing vulnerability through proactive efforts that connect with and enhance response and recovery capacity.

Community engagement

A major goal of the city's climate planning effort was to develop a plan that was supported and endorsed by all stakeholders, especially racial minorities and historically disinvested populations. Outreach and engagement was prioritised in neighborhoods where residents are most vulnerable to the current and predicted impacts from natural hazards. The engagement process addressed multiple facets of racial inequity, as well as class, disability and age discrimination.

For many people, the 'typical' city planning or town hall meeting can be incredibly intimidating. These gatherings are usually held in city buildings and include slide presentations and requests for feedback on an established plan. Community members often feel overwhelmed by the formal atmosphere or that their voices aren't heard. An important element of this type of outreach is a focus on building relationships and trust with residents, businesses and other stakeholders. Trust building takes time, empathy and patience. Many residents distrust local government for a variety of very good reasons (as mentioned earlier in this chapter). A critical element to the success of this effort was that planners took the time to actively listen to people's stories and ask follow-up questions. Additionally, with funding support from a local foundation, planners were able to remove barriers to participation by offering resources and assistance to help people attend the meetings, including free food, transportation assistance and childcare. This yearlong engagement process focused on low-income, historically disinvested communities and racial minorities (as defined in the greater US context). Additionally, elderly citizens, people with disabilities, small businesses and leaders in the nonprofit sector were also targeted.

Another crucial component of effective engagement is ensuring residents feel safe and heard. City planners held engagement meetings in venues residents identified and trusted such as community centers, churches, schools or at times, private residences. This helped improve comfort levels, which led to richer information sharing and active listening opportunities. People are more likely to listen to information from someone they know and trust (Eubanks 2015). In each neighborhood, planners partnered with a community leader when presenting information and leading group discussions. Identifying strong community leaders within each neighborhood, who encouraged residents to attend meetings and determined trusted locations for meetings, was a key to success.

In addition to outreach with residents, planners engaged with a range of stakeholders and government agencies. This helped educate decision-makers about climate change while utilising their expertise to develop the best set of strategies possible. It also cultivated a planning process that employed a diverse range of participants and provided multiple opportunities for engagement, leading to more significant buy-in, support for the plan and commitment to implementation.

A key element to engagement on climate change was to ensure that residents and stakeholders most vulnerable to hazard impacts had the opportunity to share their stories, unique neighborhood characteristics and concerns about current and future effects. In 2014, the Office of Sustainability launched a behavior-change initiative centered on preparedness and increasing community resilience. The background message of the campaign was that although mitigating of climate change through greenhouse gas emissions reductions is critical, impacts from a changing climate are already being felt throughout the world; disasters can happen any time, and we must work together to proactively prepare for these disasters to avoid negative outcomes. This message was packaged into three simple actions residents can take to be prepared: *Make a Plan, Build a Kit, Help Each Other*.

To deliver this message and engage with residents, city planners held over 40 events focused on gathering stories, actively listening to participants, developing emergency plans with residents, mapping assets in their neighborhoods and creating preparedness kits. Residents were asked to share stories about extreme weather events they experienced, and to detail the impact of those events on their homes, commutes, transport, health, daily routine and access to resources. Traditional presentation formats were intentionally avoided in favor of a more interactive approach to help build relationships and provide an opportunity for all voices to be heard.

After discussing natural hazards and history in their neighborhood, residents made emergency plans that included identifying an out-of-area contact, the nearest hospital, and additional personal information that could help in the event of an emergency. Community members then identified community assets and shortcomings. Many neighborhoods identified the need for a buddy system to help aid seniors and young children home alone after school. Others identified the need for services like a community tool-share where residents can access tools such as snow blowers to aid response to major weather events such as a snowstorm. Assets often included nearby medical centers, community gardens, positive relationships with police and fire personnel, community parks and locations with backup generators.

Last, planners worked with residents to build emergency kits. Kits include materials such as emergency radios, flashlights, can openers, first aid kits, facemasks, water storage containers and help/safe signs. Although most of these items may seem basic, the majority of residents did not own or have access to emergency materials of this nature. A unique element to Baltimore City emergency kits is the help/safe signs. Baltimore contains a large number of row houses and connected housing units. The help/safe signs provide first responders and Community Emergency Response Team (CERT) leaders with information quickly and effectively so they know where to prioritise resources. During a hazard event, residents are asked to place the sign in their window with the 'safe' sign pointing to the street if they do not need additional assistance. If assistance is needed,

Figure 8.1 City staff work with residents to identify community assets and evacuation routes

residents place the 'help' side of the sign towards the street for first responders to see. When developing kits, residents are trained on how to use each item in the kit, including the emergency radios and help/safe sign. This interactive structure keeps participants engaged, provides them with an opportunity to share their stories and helps build stronger relationships.

Since April 2014, over 2,500 emergency kits and plans have been developed with Baltimore City residents. There were several key themes identified throughout the community outreach and engagement process. Some of the main concerns identified by residents were lack of heating, cooling and ability to store backup supplies in their homes, as well as a lack of response from emergency personnel during a hazard event. Based on this feedback, a new project that would increase community resilience and adaptive capacity was born.

Resiliency hubs

In an effort to capitalise on the success of the individual preparedness campaign and provide residents with requested resources, the Office of Sustainability sought funding to develop community 'resiliency hubs'. Resiliency hubs are trusted locations for residents to gather and access resources and needed services during or after a hazard event. A resiliency hub includes a building or set of buildings that provide shelter, backup electricity, access to fresh water resources such as food, ice, charging stations, etc., in the event of an emergency. The hubs also serve as a location for incorporating energy efficiency, weatherisation, cool roofs, solar installation and battery backup systems.

Resiliency hubs enable a neighborhood or cluster of neighborhoods to adapt to climate change impacts and recover quickly from hazardous events. This project is unique in that it ties together strategies and actions within several city plans as well as implementation work, and combines them into one project to achieve the greatest impact possible. This includes increasing community equity and access to resources, improving community education and outreach, reducing heat island impacts and sequestering carbon by increasing green space and tree canopy, improving energy efficiency and reliability, and enhancing community members' ability to access resources and respond to hazard events. Resiliency hubs provide a way for the city to coordinate mitigation and adaptation efforts, some of which are already being implemented, and through organised efforts, produce a cohesive and concrete outcome that increases the adaptive capacity and resiliency of our neighborhoods and city.

The resiliency hub process began by conducting door-to-door climate surveys in the four high-priority areas. Planners collaborated with researchers at George Mason University to train students from local high schools and universities to gather baseline information from residents. This provided temporary employment for students and residents, and helped develop stronger connections between residents within their neighborhoods. The outreach process found that two-thirds of Baltimore residents surveyed (65 percent) would be somewhat or extremely likely to use a community center that provided resources during an emergency (Akerlof et al. 2016).

The resiliency hubs are intended to empower residents and provide the needed resources to prepare, respond and recover from a disaster event without assistance from the city. The four pilot sites were determined based on vulnerability to natural disasters and everyday stressors, as well as a strong commitment from community leaders. A key criterion for selecting a hub was a resident-selected trusted location that was well utilised within their neighborhoods for other activities. Planners worked with an advisory team and with community leaders to determine a set of criteria for site selection, including elements such as the condition of the building, opportunities for energy efficiency upgrades, suitability of the roof for solar with battery backup systems, storage space, surrounding lots for possible food generation opportunities, etc. Additionally, teams worked to determine phasing for each site that included elements such as conducting energy audits and roof assessments in phase one; conducting kitchen assessments and determining green infrastructure opportunities in phase two; ordering supplies and installing rooftop solar in phase three.

Each of the resiliency hubs sites is as unique as the neighborhoods they serve. Residents and community leaders were involved in every step of the development process. Planners worked with community leaders to determine retrofits needed for each building and supplies desired for each site and determine unique needs for particular sites. Each site underwent a series of assessments and upgrades. The Office of Sustainability and Mayor's Office of Emergency Management provided a series of trainings for both site staff and community members to ensure they were equipped with the appropriate tools to effectively manage a hub in the event of an emergency. To date, none of the resiliency hubs has been tested due to the project being very new and a warmer than usual winter of 2017. However, with the record-breaking heat of early 2017, they will likely be tested often throughout the summer.

Truly utilising an equity lens throughout the entire process (planning, implementation and integration) involves moving beyond extensive engagement to empowerment. The resiliency hub initiative provides an example of how an equity lens can be applied for an implementation process within specific neighborhoods. While the resiliency hub initiative was moving forward, so was the process of updating the city's sustainability plan.

Ambassador network

In 2015, the sustainability office hosted a supper with residents from all over the city. The supper was the first phase in developing a network of ambassadors interested in making Baltimore a more sustainable and resilient city. Staff and commissioners were asked to invite three people with specific consideration given to age, race and class with the intention of bringing together a group that reflected the diversity of the city. Attendees of the dinner were asked to volunteer as ambassadors and to help recruit other ambassadors throughout the city. Ambassadors are residents who volunteer to assist in conducting outreach within their communities and gathering feedback to inform the planning and implementation.

In order to reach as many neighborhoods and residents as possible, the city was divided into ten districts each with a similar number of people. Each district was assigned a team leader, who acted as the direct connection to the city and resources; two lead ambassadors, who helped organise ambassador outreach strategies and collect information; and a series of 10–15 ambassadors who lived or worked within those districts. Each district group helped identify tools and resources that would best support and facilitate effective outreach. By the spring of 2016, the ambassador network numbered 150 individuals.

Each district team was provided with trainings on race, equity and inclusion as well as sustainability and resiliency. Trainings focused on interactive opportunities to work in teams and practice interviews. In order to ensure equity was integrated into the entire process, an equity consultant was hired to improve awareness, increase understanding and build capacity to identify disparities in an inclusive and comprehensive manner. The consultant helped establish a common understanding of racial inequity and provided multiple opportunities through meetings and trainings to communicate and educate everyone involved in the process about racial inequity. The trainings also provided the entire ambassador

Figure 8.2 Baltimore City ambassador training in the summer of 2016

network with the opportunity to help shape the outreach toolkits, identify incentives they could hand out to facilitate engagement and develop survey questions that guided the outreach process.

Although there was very little funding for the outreach process, it was essential to compensate volunteers as much as possible. Lead ambassadors were provided with a small stipend for their leadership role and data collection requirements. Each district was given a gift card worth $300 to use however they felt would best support their outreach process. Districts chose to use their gift cards for events such as block parties or for purchasing neighborhood-specific incentives. This approach gave decision-making power to each district, and allowed ambassadors to determine what would best support their outreach process.

In addition to the ambassador network, project teams were developed to help direct the process and conduct outreach with special interest groups. Specific consideration was given to race when developing the teams, with the goal of 65 percent African American representation on each team from a range of neighborhoods. These internal project teams helped guide and revise the outreach strategy, support the structure of the plan and provide input and feedback throughout the process.

Intentional engagement and outreach was also conducted with youth and special interest groups. A youth working group and youth ambassador network, led by high school interns, led the youth engagement process, which included hosting their own town hall event and conducting targeted outreach. In addition, over 25 meetings were held with institutions and special interest groups including environmental justice and civil rights organisations. These meetings helped inform stakeholder groups about the planning process, solicited feedback, obtained commitments for ongoing collaboration in implementation phases and helped strengthen relationships.

Communications

Communication is a key ingredient to building trust and facilitating strong relationships. Climate change, resiliency and sustainability are all broad concepts that are often difficult to explain or understand. The Sustainability Office sought to simplify these topics by making direct connections to the amazing work members of the community are doing that helped make Baltimore a more resilient and sustainable city. In 2016, the *Every Story Counts* campaign was developed. The campaign captures Baltimore City residents' stories through their own words and with one simple picture. For example, in early 2017, the campaign featured four amazing women who help provide support to residents who don't speak English and need help navigating different systems, including healthcare, transportation and/or bill paying. It demonstrates how their efforts build stronger relationships, which also enhances community adaptive capacity. The *Every Story Counts* campaign focuses on empowerment and recognition of residents' actions over government actions. It also addresses hard issues like drugs, poverty and incarceration and connects the work being done to improve those issues with enhancing resiliency and sustainability.

Figure 8.3 Baltimore resident Robbyn Lewis's Every Story Counts profile picture. Robbyn and her neighbors created their alley mural to help prevent illegal dumping.

Training

In addition to strong communications campaigns and relationship building, providing effective training is a critical element to enhancing resilience. This includes trainings for residents and stakeholders (detailed earlier in this chapter) as well as for city staff. While opportunities to address climate change are abundant, city staff tends to focus on getting day-to-day work accomplished, with little reference point for how climate may be affecting their work or where they may have opportunities to incorporate adaptation. Successful climate adaptation requires consideration of future climate conditions across city functions from land use planning to infrastructure design and maintenance to emergency management.

In 2016 the Baltimore Office of Sustainability received a grant to develop a toolkit that utilised real-world examples to help develop customizable training materials for city employees. After working with a consultant team and other city representatives, the outcome was two interactive activities (games) structured around the vulnerability assessment process. The process focused on flooding from precipitation, tidal flooding and sea level rise and led to the development of *The Game of Floods*, a team activity that focuses on collaboration in identifying assets, determining vulnerability and prioritising actions to protect people and property from impacts given a specific climate scenario and budget. These interactive trainings help city staff better understand how climate adaptation integrates into their daily work and decision-making.

Figure 8.4 USDN Game of Floods training for sustainability directors, engineers and planners in 2017

Implementation efforts

Implementation is critical to the success of any planning process. In order for there to be equitable implementation, stakeholders and residents must have easy-to-access feedback opportunities and continuously be engaged as partners in the process. In Baltimore, projects also need to be focused in areas without equal access to resources and most vulnerable to shocks and stresses. Due to the comprehensiveness of the City's DP3, priority actions include mitigation and adaptation co-benefits and are focused in areas most vulnerable to impacts from climate change. Since adoption of the plan in 2013, Baltimore City has successfully implemented a large number of strategies and actions. As a way to highlight some of these without detracting from the main content of this chapter, they are listed in the table below.

Conclusion

Although the planning and implementation processes discussed in this chapter made genuine attempts at utilising equity as a lens, they still had shortcomings. It was very difficult to keep participants engaged throughout the entire process, especially people who were volunteering their time, such as ambassadors. The Office of Sustainability could have done a better job at continuous support and engagement with the ambassador network. The ambassador network experienced a significant drop-off after the equity and sustainability trainings were complete. Ambassadors were supposed to reach out to a minimum of five people over a two-month period and solicit their feedback. Based on comments gathered mid-way into the process, ambassadors who dropped out or didn't complete their outreach admitted to not prioritising outreach efforts over other family commitments. Additionally, districts with strong lead ambassadors and strong outreach plans were more successful at ambassador retention. Strong lead ambassadors were key to effective outreach and engagement.

Efforts were also made to have committees, ambassador teams and internal working groups represent the socio-economic makeup of Baltimore. However, that target was difficult to achieve. In addition, more direct efforts were needed to ensure that not only race was considered but also economic status, age, ability level, etc.

In addition to challenges around equity, climate change and integrating climate into everyday decision-making is also extremely difficult. Baltimore City faces a wide range of stressors such as epidemic drug use, high crime and poverty. Most people, including decision-makers, are unable to see proactive planning for climate change as an important issue compared to the stressors they see every day. In fact, typically, until a catastrophic event hits home, very little attention or funding is given to natural hazard mitigation or climate adaptation. A struggle in this process has been making the connection for stakeholders and decision-makers between daily stressors and how those stressors will continue to be exacerbated by a changing climate. It's a challenge to show the benefit of taking proactive action now to prevent likely future disaster. Although high-profile

Table 8.1 Baltimore City climate & resilience implementation projects

Action	*Main content*	*Equity consideration*
Update of the City's floodplain code	The code included adoption of higher regulatory standards such as development of a Flood Resilience Area, extending regulation to the 0.2 percent chance flood, increasing the city's freeboard to 2-feet, removing exemptions for historic properties and incorporation of ASCE-24 into the code.	These updates proactively protect new and retrofitted structures that are vulnerable to flooding and SLR while also reducing risk to people and property.
Community Rating System (CRS) Certification	The Community Rating System (CRS) is a voluntary incentive program that encourages community floodplain management activities that exceed National Flood Insurance Program (NFIP) requirements. Proactive measures around public information, mapping and regulations, flood damage reduction and flood preparedness. In May 2015, Baltimore was awarded a Class 5 CRS ranking providing policyholders with an 25 percent discount in their flood insurance rates.	As an incentive, flood insurance premium rates are discounted for all policyholders to reflect reduced risk. This lowered flood insurance costs for residents and businesses.
Tree plantings and impervious surface removal	The city has been working with a number of non-profit partners to help train residents and business owners to plant and properly care for trees. The city's Tree Keeper program was launched in 2013 and includes several levels and types of educational and hands-on classes that certify residents to perform work on public trees. Additionally, the city collaborates on impervious surface removal to help improve water quality and quantity management.	Tree plantings are targeted in areas with high impervious surface and low tree canopy. This helps reduce urban heat island effect and improve air quality.

Adoption of the International Green Construction Code (IgCC)	The International green Construction Code (IgCC), adopted in 2015, is the first model code to include sustainability measures for an entire construction site and process including design, construction, certificate of occupancy, and beyond.	This will make buildings more efficient and improve health, safety and community welfare.
Growing Green Initiative (GGI) and Green Network Plan (GNP)	The GGI uses sustainable, innovative, and cost-effective practices for reusing vacant land to green neighborhoods, reduce stormwater runoff, grow food, and create community spaces. The City is now working on development of a GNP to connect these green spaces and better link neighborhoods.	Priority areas include low-income neighborhoods with communities of color and areas vulnerable to high heat.
Integration of resilience into the City's Capital Improvement Process (CIP)	A capital improvement is a long-term investment, typically in physical infrastructure, which improves the city. Each city agency must go through a checklist of qualifications and requirements when submitting requested funding for a capital improvement. In 2014, Baltimore City incorporated a climate and resilience checklist into the CIP that ensures city agencies are considering climate change in their long-term planning and funding requests.	This ensures that other city departments are considering climate change and areas most vulnerable to impacts when seeking funding and support for projects.
Star Community Rating System	The STAR rating system provides a data-driven approach to assessing communities' sustainability efforts. This helps Baltimore track progress toward overall sustainability objectives and compare itself with other cities. In 2015, Baltimore City was recognised as a top achiever in national sustainability and was awarded a STAR-5 rating.	Baltimore was able to identify areas for improvement. Equity and Empowerment is a major area the City is now focusing its efforts.

events like Hurricane Sandy and Hurricane Katrina help to reference, until a large event takes place in Baltimore, many people still see climate change as something that will happen in the future rather than something that is impacting their quality of life now.

Although there are many contending priorities and needs at a variety of scales, Baltimore City has taken great strides in climate adaptation planning and implementation over the past few years. Utilising equity as the spine for ongoing engagement and implementation has prioritised resources, strengthened community adaptive capacity and improved relationships with residents. There is still a tremendous amount of work to do in both reducing risk to impacts from climate change and becoming a more equitable city. Political support, strong leadership, dedicated multi-year funding and building relationships are critical components to proactively strengthen neighborhoods in an equitable and just way. Effective climate change planning and implementation requires all stakeholders to address historic injustices and prioritise solutions that commit resources in ways that are just and fair; in ways that remove barriers and providing the resources people need to be resilient and lead healthy and full lives.

As stated in the introduction of this chapter, building resiliency starts with acknowledging current risks while focusing on reducing vulnerabilities. However, it must also include acknowledgement of past injustices and anticipation of future impacts. For low-income residents and communities of color, there is a need for increased capacity to overcome additional stresses that may result from a disaster. This includes enhancing the community members' access to resources such as communications, medical supplies, food, water, heating and cooling. In addition to resources that aid in response and recovery from hazard events, there are other more beneficial resources and initiatives, such as increased community green space and the installation of renewable energy systems with backup generators, which assist in proactive disaster preparedness and increasing overall community resilience.

Bibliography

Akerlof, K., Dindinger, J., and Rowan, K. (2016). *Perceptions of Community Resilience: A Maryland Community Pilot Study, 2016*. Fairfax, VA: Center for Climate Change Communication, George Mason University.

Baltimore Development Corporation (2017). Retrieved January 15, 2017 from, http://baltimoredevelopment.com/why-baltimore/economic-profile/

Boon, J., Brubaker, J.M., and Forrest, D.R. (2010). *Chesapeake Bay Land Subsidence and Sea Level Change an Evaluation of Past and Present Trends and Future Outlook*. A report to the U.S. Army Corps of Engineers Norfolk District. Special Report No. 425 in Applied Marine Science and Ocean Engineering. Virginia Institute of Marine Science (VIMS).

Bradley, R., Karmalkar, A., and Woods, K. (2010). *Climate System Research Center (CSRC)*. Amherst: University of Massachusetts Amherst.

City of Baltimore (2016). *Data Source*. Baltimore: Department of Public Works.

Cooley, H., Moore, E., Heberger, M., and Allen, L. (2012). *Social Vulnerabilty to Climate Change in California*. Oakland: Pacific Institute.

Cronin, T.M. (2013). *Science Summary – Sea-Level Rise and Chesapeake Bay. U.S. Geological Survey*. Retrieved from https://chesapeake.usgs.gov/sciencesummary-sealevelrise.html. (2017.01.10).

Davoudi, S. (2012). Resilience: A bridging concept or a dead end? *Planning Theory & Practice*, 13(2), 299–307.

Environment Maryland Research & Policy Center (2012). *When It Rains, It Pours: Global Warming and the Increase in Extreme Precipitation from 1948 to 2011*. Retrieved from www.environmentmaryland.org (2015.11.26).

Eubanks, P. (2015). The Troubled Rhetoric and Communication of Climate Change: The Argumentative Situation. London and New York: Routledge.

Jerrett, M., Su, J.G., Reid, C.E., Jessdale, B., Ortega Hinojosa, A.M., Shonkoff, S.B., Seto, E., and Morello-Frosch, R. (2012). *Mapping Climate Change Exposures, Vulnerabilities and Adaptation to Public Health Risks in the San Francisco Bar and Fresno Area*. Berkeley: The University of California.

Larsen, C.E. (1998). *The Chesapeake Bay: Geologic Product of Rising Sea Level. U.S. Geological Survey*, October. USGS Fact Sheet 102–98.

Maryland State Archives (2017). Retrieved February 20, 2017 from, http://msa.maryland.gov/msa/mdmanual/01glance/html/port.html

Morello-Frosch, R., Pastor, M., Sadd, J., and Shonkoff, S.B. (2008). *The Climate Gap: Inequalities in How Climate Change Hurts Americans & How to Close the Gap*. Los Angeles: PERE.

National Climate Assessment- heavy precipitation in the Northeast United States (2014). Retrieved December 2016 from, http://nca2014.globalchange.gov/report/regions/northeast

NOOA (2013a). *Tides and Currents- Tides and Water Levels. Mean Sea Level Trend 8574680*. Baltimore, MA: National Oceanic and Atmospheric Association (NOOA).

NOAA (2013b). *Regional Climate Trends and Scenarios for the U.S. National Climate Assessment: Part 1: Climate of the Northeast U.S.* National Oceanic and Atmospheric Association (NOAA) Technical Report NESDIS 142–1.

Porter, L., and Davoudi, S. (2012). The politics of resilience for planning: A cautionary note. *Planning Theory & Practice*, 13(2), 329–333.

Pulido, L. (2000). Rethinking environmental racism: White privilege and urban development in Southern California. *Annals of the Association of American Geographers*, 90(1), 12–40.

Reeves, D. (2005). *Planning for Diversity: Policy and Planning in a World of Difference*. London: Routledge.

Resilience Alliance (2007). Assessing and managing resilience in social-ecological systems: A practitioners workbook. *Resilience Alliance*. Retrieved from www.resalliance.org

Spanger-Siegfried, E., Fitzpatrick, M.F., and Dahl, K. (2014). *Encroaching Tides: How Sea Level Rise and Tidal Flooding Threaten U.S. East and Gulf Coast Communities Over the Next 30 years*. Cambridge: Union of Concerned Scientists.

University of Maryland Center for Environmental Studies (2014) State Report.

USCB (2010). *Census. U.S. Census Bureau*, December 10.

USGS (2015). *Analysis of Storm-Tide Impacts from Hurricane Sandy in New York*. United States Geological Survey. Scientific Investigations Report 2015–5036. Retrieved from https://pubs.usgs.gov/sir/2015/5036/sir20155036.pdf (2016.12.28).

Walsh, J., Wuebbles, D., Hayhoe, K., Kossin, J., Kunkel, K., Stephens, G., . . . Somerville, R. (2014). Our changing climate: Climate change impacts in the United States. The third national assessment. In: Melillo, J., Richmond, T.C., and Yohe, G.W. (eds.) *National Climate Assessment Report*. Washington, DC: U.S. Global Change Research Program, 19–67.

Wilkinson, C. (2012). Urban resilience: What does it mean in planning practice? *Planning Theory & Practice*, 13(2), 319–324.

9 Towards transformative action

Learning from local experiences and contexts

Susie Moloney, Hartmut Fünfgeld and Mikael Granberg

Introduction

In this book, we have brought together a range of experiences and examples of local climate change responses and actions from diverse contexts around the world, including Australia, Vanuatu, South Africa, Japan, India, Sweden and the USA. This collection of case studies on local climate change action offer insights into the range of complex social, institutional and political challenges and issues mediating the work of local actors as they grapple with climate change. The examples also highlight the many ways in which local and regional context matters and how particular individuals and social groups emerge as key actors progressing climate change action – often through collaboration among different social groups from across local government, private sector and civil society. While in many cases the actions being adopted have similarities – including risk assessments, goal setting, strategic planning, project implementation, emergency management, etc. – the experience of 'doing' climate change mitigation and adaptation is very much shaped and influenced by the different historical, geographical, socio-political and cultural contexts in which they take place. In examining different local and regional contexts, we were interested in understanding how local contingencies and specificities create conditions that can both enable and constraint innovative and transformative actions – those actions that have the capacity or potential to challenge or change business-as-usual planning and decision-making processes and their outcomes and more broadly in terms of governance, policy and practice change. In focusing on local government actors and those they collaborate with, these specific contextual conditions manifest in how climate change is interpreted, how problems are framed and prioritised, how policy responses emerge, and in relation to the capacity to innovate and experiment new ways of doing – organisationally, institutionally and socially.

The case studies also point to the inherent constraints to acting locally on climate change – particularly in terms of the capacity and power of local scale responses to effect transformational change. Despite these challenges – or perhaps because of them – there is growing interest in developing local actions and capacities, as many of the case studies discussed in this book illustrate. Developing capacity and creating and using opportunities for transformative change rely on different forms of knowledge to inform key stages in decision making – both with regard to understanding the likelihood and severity of climate change impacts (i.e., flooding events) and how different actors (policy makers, communities, households) can effectively plan and respond in the short, medium and long term. Alternative forms of knowledge creation and use, such as tapping into tacit or explicit local knowledge, co-creating new contextual knowledge through collaboration of scientific information with traditional forms of knowing, can offer a creative bed of ideas, from which alternative options for climate change action and transformational processes can flourish and grow. While national or top-down plans, strategies and resources are important, local capacities to understand, shape, enact and implement actions are critical for devising locally adapted solutions that are well-suited to context, effective in achieving set goals and supported by local constituencies.

Attention then turns to understanding the factors enabling and constraining local capacities in its multiple forms – socially, organisationally and institutionally – and how different actors work both within existing constraints and how they seek to build capacities to engender more innovative and transformative responses and outcomes. While focusing on the local scale, we must acknowledge that climate change responses are conditioned across multiple scales and institutions, involving different levels and modes of governance. The authors situate each case study within their local and multi-scalar context, highlighting particular histories, politics and social relations motivating and shaping actions. Issues of inclusion and exclusion are identified in many instances, in terms of different actors' involvement and capacity to shape climate change actions and responses. The case study authors explored how dominant policy frames and logics enable climate change responses – or act as constraints on innovation. The use of, and need for, different forms of knowledge and the question of how knowledge is created and managed also emerge as key issues. A key theme across many cases is the necessity for local governments to work in new, more collaborative ways to build or expand their capacity and power to influence decision-making in their local constituencies.

As a way of synthesising the case study findings, the following discussion includes a brief descriptive review of each case, followed by an analysis of the conditions enabling and constraining innovative and transformative action. By learning from practice in a range of local contexts, we aim at advancing both our empirical and theoretical understandings of local climate change action.

Learning from local contexts and experiences

Greater Geelong and Melbourne, Australia

In Chapter 2, *Collaborative governance: a means for integrating climate change action in Australian local government?*, Hartmut Fünfgeld and Susie Moloney explored the ways in which local governments in the municipalities of Greater Geelong and Melbourne are building their capacity through trialling innovative organisational processes and collaborative forms of governance in the form of intra and inter-municipal collaborations. The chapter highlighted both the necessity to work through partnerships and the challenges involved in governing collaboratively through informal/formal processes. While demonstrating the important catalyst and locally coordinating role of local governments in climate change planning, the Australian case studies also showed that mainstreaming climate change across local government department and practices is an ongoing challenge.

Port Vila, Vanuatu

In chapter 3, *Governance and agency beyond boundaries: climate resilience in Port Vila's peri-urban settlements*, Alexi Trundle reviewed urbanisation processes in the context of climate resilience in the peri-urban areas of Port Vila, Vanuatu. The analysis highlighted the importance and value of including indigenous communities and knowledge in formal urban planning processes. It illustrated both the challenges with and the benefits arising from understanding the locally specific geographical, historical and cultural context within which peri-urban development is planned and managed, including constraints associated with arbitrarily drawn urban boundaries in managing climate change adaptation, climate resilience and vulnerabilities.

Western Cape Province of South Africa

In chapter 4, *'Learning by doing': lessons from the co-production of three South African municipal climate change adaptation plans*, Julia Davies and Gina Ziervogel examined the process of developing climate change adaptation plans in three local settings in the Western Cape Province of South Africa. Three case studies highlighted the need for tailor-made, locally specific approaches to local climate change adaptation. Political will and strong support from senior public officials was shown as critical to mobilising adaptation plans, along with employing a collaborative and integrated approach in enabling more effective policy processes. In the South African examples, the urgency and priority given to climate change, however, was shown to be mediated by more pressing social and economic challenges, which present challenges for mainstreaming climate change adaptation into local decision-making.

Rural and regional Japan

In chapter 5, *Facilitators and barriers: regional diversity among local climate protection projects and regional revitalisation efforts in rural Japan*, Christian Dimmer and Daniel Kremers presented two cases of local climate action in rural and peripheral Japan. They situated these responses within the context of pressing socio-economic challenges – rapid demographic ageing and a high level of energy dependency and insecurity. Through this lens, they examined the relationship between climate action and the prosperity of two local communities. From the two case studies discussed, the role of national-level goals and plans present unique challenges in terms of implementation at the local scale, where municipalities can lack capacity, vision or knowledge to enact national objectives. This chapter again illustrated in poignant ways the mediating role of other key policy objectives in responding to climate change and the value of emphasising co-benefits to achieve broader economic and social objectives and working towards improving local conditions in ways that can also improve community resilience and adaptive capacity.

Bangalore and Chennai, India

In chapter 6, *Multi-level climate change planning: scale, capacity, and the authority for local action*, Neha Sami examined local climate change planning in two Indian cities, Bangalore and Chennai. Here, the focus was on urban settlements and the ability of urban policy makers to respond to climate-related challenges. While national-level directives are clearly motivating and framing regional and local-scale action, the Indian case studies demonstrate how important and at the same time, how difficult it is to ensure multi-scalar coordination of climate change policies and action across different agencies. The roles and responsibilities for urban planning and governance are distributed across a range of city, parastatal and state-level agencies. While national objectives may provide some framework for action, the lack of coordination across scales, lack of capacity and willingness to prioritise climate change in urban development policy making is significantly constraining transformative potential.

Karlstad, Sweden

In chapter 7, *Climate change adaptation, city competitiveness and urban planning in the city of Karlstad, Sweden*, Mikael Granberg and Lars Nyberg presented the case of municipal climate change adaptation planning in Karlstad, Sweden. They highlighted innovative processes adopted by local actors in engaging with the challenges of adaptation while, at the same time, acknowledging the constraints of dominant policy priorities shaping city development goals and processes. The Karlstad case study brought to the fore the inherent tension between 'business-as-usual' city growth and competition goals focused on waterfront housing developments on the one hand and the need to manage climate-related risks related to future inundation on the other hand. While Sweden is demonstrably a global leader in environmental sustainability and climate change policy making, the dominance of a growth imperative continues to frame local climate change policy and action and as in the case of Karlstad, outweighs the potential impacts of climate change risks.

Baltimore, USA

In chapter 8, *Climate Resilience: Transformative Efforts in Baltimore City*, Kristin Baja and Mikael Granberg examined local government climate action, focusing on innovative and proactive planning around risk reduction and preparedness in the city of Baltimore in the United States. The Baltimore case provided an interesting example for making sense of and working towards increased climate change resilience in a racially segregated city. The city's history and social and economic context placed equity considerations, in dealing with areas and communities most vulnerable to climate change, at the forefront of transformative thinking. Here, if planning for and implementing climate change actions is to truly build social resilience, planning and decision-making processes must address issues of social exclusion, racial inequality, lack of resources and accessibility. The Baltimore case demonstrated the important and innovative role of the city government and its commitment to addressing many of the profound social, economic and climate-related challenges in the most vulnerable communities.

Conditions enabling and constraining innovative and transformative action

In Chapter 1 we provided an overview of key dimensions of climate change actions as they relate to the local and regional scale, as a framework for situating local actions and practices in responding to climate change (see Table 1.1 in Chapter 1). We also developed a framework for interpreting and reflecting on conditions that enable and constrain innovative and transformative action at the local scale. In Table 9.1, below, we revisit this framework and, by way of synthesising findings emerging from the case studies, identify conditions that either have the potential to enable or constrain innovative and transformative action on local climate change mitigation and adaptation. While offering a useful heuristic device, we recognise that this framework and the conditions identified are not reflective of the complexity of possible conditions enabling and constraining local climate change action. This framework however provides a lens through which key issues and lessons informing and shaping different types of local actions can be interpreted, and to examine the extent to which local conditions are enabling or constraining the potential for innovative and transformative climate change action.

1. *Clarity of roles and responsibilities: weak role of local government constraining capacities*

Drawing on the case studies presented in this volume, we can discern conditions enabling and constraining local transformative potential and highlight lessons that can inform climate planning and action in other contexts. Unsurprisingly, a critical issue with regard to the potential to impact the capacity and efficiency of local climate change action is the formal capacity and decision-making power of local government (Fünfgeld & Moloney, Davies and Ziervogel, Sami, Granberg & Nyberg). In a weak local government system, there is a need for local public actors and agencies to find resources outside the public sector to facilitate local climate action capacity. Similarly, in contexts of relative strength or organisational leadership, local government's role in influencing external actors presents

Table 9.1 Conditions enabling and constraining local innovative and transformative action on climate change mitigation and adaptation

Conditions	*Institutional and organisational dimensions in which condition applies*	*Enabling condition*	*Constraining condition*
1. Clarity of roles and responsibilities	Multi-level governance of climate change Intra-organisational division of roles and responsibilities	Efficient and effective planning and decision-making across multiple scales and institutional levels	Confusion about roles and responsibilities leading to stalling and inaction Conflicting actions increasing emissions and maladaptation
2. Political commitment and high level support	Democratic decision making processes Organisational buy-in	Political commitment demonstrates importance and enables buy in from across the organisation/ constituency	Lack of political commitment leads to neglect or active discouragement of climate change action
3. Organisational and institutional co-ordination and integration	Horizontal and vertical collaboration across institutions and departments	Well co-ordinated climate change action leads to mainstreaming into all organisational processes	Poor coordination leads to sporadic and fragmented action
4. Problem/policy framing and formulation	Multi-level/actor participatory approach Value based framing and prioritisation in the formulation of issues on the policy agenda	Climate change is framed as an all-encompassing problem that requires coordinated and sustained action	Climate change action remains a niche task of environment departments
5. Local context and specific circumstances	Translation of climate challenges/ problems in relation to local contextual conditions Inclusion of multiple local scale actors and forms of knowledge in decision making	Climate change action is tailored to and grounded in locally specific circumstances, drawing on local knowledge	Climate change action ignores local knowledge and context and is devised in a top-down fashion
6. Social learning and experimenting	Developing capacity through breaking path dependency Commitment to innovation, reflexive learning and developing new pathways of action	Experimenting and social learning is encouraged, leading to innovation and transformative action	Experimentation is discouraged or not resourced and climate change action is required to fit into 'business as usual'

an opportunity to effect broader transformation and more coordinated responses. Hence, mobilising additional, externally held resources is an important driver for local government collaboration with external actors through networked forms of governance. Collaborative forms of network governance can facilitate local capacity building and more inclusive and creative governance models for handling complex societal challenges, like climate change, in an interconnected world.

In terms of multi-level governance, issues on the local policy agenda also impact the implementation of national climate action policies and decisions. National policy implementation has to be understood in relation to how other major challenges are perceived and prioritised in the specific local setting. Ambitious objectives formulated by higher-tier government risk running into structural obstacles hindering implementation at the local level when local actors and local government lack capacity (knowledge or resources) and/or interest to implement national objectives. It is vital that national government frame local climate action in ways that emphasise the co-benefits of local action. More consideration is needed around the ways local climate action can realise other goals that can improve local conditions in ways that also increase adaptive capacity and community resilience.

2. *Political commitment and high-level support*

For local climate action in general, but in particular for developing governance structures and deciding on concrete action on climate change, political and senior management level support and acceptance of climate change is critical (Davies and Ziervogel, Granberg and Nyberg, Baja and Granberg). Fundamental understanding of the challenges at hand, acceptance of anthropogenic climate change and support for local action enables the central positioning of climate issues on the political agenda and directly influences the allocation of resources (money, competencies, etc.) as part of the political process of local decision-making. Political will and strong support from senior public officials for the climate action agenda, paired with an integrated approach to climate change planning, are important for enabling more effective policy responses to climate change.

Relations between the state and local government is of course relevant when discussing political and higher-level support for local climate action. National governments set objectives but also make decisions (allocation of resources, etc.) that impact the conditions for local actors, and local government in particular, in terms of regulation, support and control through national authorities and agencies. Our cases show considerable variation in this regard. Sweden (Granberg and Nyberg) stands out as an example of strong national government using regulation and policy decisions and allocating resources to local government climate action but at the same time giving considerable autonomy to local actors to translate action into specific local contexts. India (Sami) and Japan (Dimmer and Kremers) clearly illustrate the classic implementation challenge where national government make policy decisions that is not fully implemented at the local level due to lack of local capacity or due to a perceived mismatch between national decisions and the needs and challenges in a specific local context.

3. *Organisational and institutional collaboration, co-ordination and integration*

Developing innovative processes to improve organisational and institutional collaboration, co-ordination and integration is central to the challenge of governing climate change. While we are seeing new forms of organisational decision-making and institution building, further research is needed to understand how effective these are at dealing with complexity. Intra-municipal and inter-municipal collaboration is emerging as an important feature of local climate governance and warrants further exploration around how these processes work and what challenges are confronted in mainstreaming and embedding climate change across all areas of decision making (Fünfgeld and Moloney, Davies and Ziervogel, Granberg and Nyberg, Baja and Granberg). Collaboration can enhance capacity but can also create new challenges around navigating informal/formal governance processes and how lessons from different forms of experimentation (i.e., new projects or capacity building processes) can influence and change existing practices and formal decision-making. From an instrumental 'power to' perspective, network governance can be helpful for enhancing local climate action. From a more critical power perspective, however, there is the potential for some forms of network governance to have limited transparency with regard to decision-making processes and democratic accountability. Issues of utilisation of knowledge, who is included and excluded in the decision-making process and how this impacts social priorities in local climate action needs to be brought to the fore and critically examined.

4. *Problem/policy framing and formulation: tensions and co-benefits*

A key challenge in prioritising and mainstreaming climate change adaptation is how it is defined and framed (conflict or complementarity) with other pressing policy issues on the local policy agenda. Local climate action can be mediated through dominant policy priorities shaping wider city development goals. For instance, many local governments around the world perceive that they need to make their city more competitive. They believe they have to be active and innovative to increase competitiveness and this can create a frame within which formulation and priority of climate actions is defined. There is a significant risk that if the competiveness imperative dominates, it will outweigh the potential impacts of climate change risks limiting the frequency and the scope of local climate action.

In this way, problem framing and subsequent policy formulation can enable and constrain problem-solving capacity within the area of climate change, and accordingly, it is of central importance to understand how local climate action is interpreted, framed, formulated and prioritised. As touched on above, it is not self-evident that issues of climate change will end up high on the policy agenda in different local settings. In order to be utilised in local practice, climate issues need to be given a place and a priority on the local political/policy agenda. Many issues compete for importance and priority on local policy agendas, and climate action is one of these issues. If climate change is perceived as a highly important issue, it can be given priority in its own right. If not, it needs to complement and reinforce other prioritised issues to be able to compete on the policy agenda.

The framing of climate change action as a policy issue impacts how it is addressed in a specific policy context. For instance, the capacity and power to influence 'climate compatible' urban planning decision-making is critical to ensuring that local climate action is integrated into development outcomes (Fünfgeld and Moloney, Trundle, Davies and Ziervogel, Sami, Granberg and Nyberg). Accordingly, the policy formulation of climate change, and its relation to other highlighted issues, visions and objectives, is of great importance for the perception and handling of the climate change challenge. Hence, addressing climate change will always be a contested process as it is an issue among others on the local policy agenda consisting of competing priorities and policy goals that may only be possible to satisfy at the expense of each other. Several of the cases in this volume illustrate this particular challenge.

5. *Local context and specific circumstances: history, politics, culture, knowledge*

Understanding local climate action and the capacity to achieve particular outcomes requires a good appreciation of how forms of action are situated in an historical understanding of the social, environmental and economic history of the specific place being examined (Trundle, Baja and Granberg). In focusing on the local scale, we recognise that success and failure of local climate action is very much shaped by place-based specificities. Specific contextual factors have emerged over time and influence the structures of power, capacities and vulnerabilities in particular local settings. The socio-economic makeup of a city or municipality is developed over time and has significant impact on the ability and capacity of local actors to act on climate change and to withstand or adapt to its impacts. The cases presented here incorporate a wide range of socio-economic and culturally diverse contexts globally and within their national and regional contexts. For example, many cities around the world, and among the case studies in this volume, are racially and culturally segregated (Trundle, Davies and Ziervogel, Baja and Granberg). Addressing climate change and enabling climate resilience through equity goals, for instance, therefore requires an inclusive approach where understandings of history, cultural diversity and social inequality underpin decision-making and that those communities most vulnerable receive the necessary resources.

Enabling climate-resilient local communities therefore means that social equity issues must be central to climate change action and to ensure that action does not further exacerbate social division and increase vulnerabilities for disadvantaged groups. The history of segregation in many places around the world is directly related to a range of contemporary economic, social and ecological challenges. Climate action, rather than being understood as a separate or unrelated challenge, needs to be instrumental in creating climate resilience in a complex social setting, combatting both impacts of climate change and social injustices (i.e., climate justice). Accordingly, proactive and innovative local government action is vital in addressing social, economic and environmental impacts in a context where certain communities and residents are disproportionately affected by the impacts of climate change.

Recognition and consideration of place based social relations and challenges raises the issue of how to navigate and include different forms of knowledge, informal and formal processes and understandings of place (Trundle, Baja and Granberg). Enabling the inclusion and empowerment of indigenous communities, as highlighted in the Port Vila case, involves creating conditions for facilitating social learning and integrating experiential (non-expert) knowledge. This reinforces the need for an in-depth understanding of the historical and cultural origins of place and of the locally specific informal arrangements they maintain and sustain.

6. *Social learning: innovation through experimentation*

It is clear from some of the studied cases in this volume that local climate action is very much a new activity for some local governments and that as such local capacity for climate action is only emerging (Davies and Ziervogel, Sami). As there is no transferrable 'one-size-fits-all' approach to responding to specific local climate change issues, local governments are compelled to adopt a 'learning-by-doing' approach to climate change adaptation which takes time and resources and can affect the efficiency of measures utilised. Lack of experience and easily applicable frameworks mean that local governments are using experimental approaches to local climate action and capacity building. Such experimentation, while fraught with logistical and legitimacy challenges in settings of constrained public budgets and ex ante demonstration of hard, scientific evidence, can also lead to the emergence of more radical transformative action that challenges the status quo and highlights the merits of alternative approaches.

Finally: knowledge gaps and issues for future research

While the case studies provided important insights into many of the conditions enabling and constraining local climate change action, they invariably only touched on particular challenges that warrant further in-depth analysis. Further research is required to better understand the interactions between different actors, multiple layers of decision-making and geographically and institutionally connected constituencies, within and beyond the local scale. Specifically, the case studies highlighted the need for further action in the following four areas:

1 Research on the political, broader institutional and social dimensions of how to create effective multi-level governance for enabling locally effective and appropriate climate change action. Here, the challenge is to come up with evidence and transferable approaches for how roles and responsibilities of

different scales of government are best distributed and shared. This requires obtaining an intricate understanding of the administrative and political workings of the broader political system and in particular, its bureaucracy in different contexts. Calls for horizontal and vertical integration need to be underpinned by realistic assessments of common obstacles and strategies for overcoming these – in particular also the potential for truly transformative action in otherwise largely inflexible government structures that are driven by political interests rather than a desire to deliver ground-breaking and most effective responses to emergent and hitherto poorly defined policy problems.

2 Critical appraisals of different mechanisms for devising and implementing innovative climate change action that go beyond high-level strategising. This includes realistic assessments of financing models for innovative mitigation and adaptation actions, new processes for ensuring existing social vulnerabilities are reduced rather than exacerbated as part of climate change actions, and also the question of what constitutes innovation and transformation in the context of dominant policy drivers, such as growth-based economic models.

3 Research into discursive and deliberative approaches for planning and decision-making that help break down internal silos, enable collaboration across administrative boundaries and allow for tapping into the creative and transformative potential of interdisciplinary and transdisciplinary action. Given the context specificity of, and therefore the lack of ready-to-apply models for, effective climate change action, the emphasis here needs to be on creative and constitutive processes that allow for the emergence of radically different ideas that can then be tested in real contexts, e.g., through urban laboratories.

4 Collaborative work to increase the capacity of local actors to learn through reflexive processes (i.e., social learning, monitoring and evaluating (M & E)) to improve ongoing climate change actions and decision making and embedding lessons that can effectively inform future actions. While requiring multi-level collaborations, this also involves developing M & E processes at the early design phase of policy, processes and projects to ensure there is commitment to ongoing learning. While there are a growing number of frameworks for M & E focusing on climate change adaptation, many have not been developed or implemented at the local scale. Further research is required around appropriate frameworks, indicators and learning models for critically assessing and tracking progress on climate change actions, changing vulnerabilities and adaptive capacity building across a range of actors. This raises issues around access and availability of different forms of data and the capacity and competency gaps needed to enable effective decision making. While M & E is important in assessing effective and efficient actions and outcomes, what must be emphasised is the use of M & E to learn about effective implementation processes and how new modes of governance and experimentation are improving climate change outcomes.

Index

Page numbers in italics indicate a figure and page numbers in bold indicate a table.

For Product Safety Concerns and Information please contact our
EU representative GPSR@taylorandfrancis.com Taylor & Francis
Verlag GmbH, Kaufingerstraße 24, 80331 München, Germany